2nd International Symposium on Materials Chemistry (ISyMC 2020)

IOP Conference Series: Materials Science and Engineering
Volume 1204

Boumerdes, Algeria
5 – 7 October 2020

Editors:

Souad Mesdour-Hammadou
Ali-Nordine Leklou

ISBN: 978-1-7138-4663-5
ISSN: 1757-8981

Printed from e-media with permission by:

Curran Associates, Inc.
57 Morehouse Lane
Red Hook, NY 12571

Some format issues inherent in the e-media version may also appear in this print version.

Institute of Physics
Dirac House, Temple Back
Bristol BS1 6BE UK

Phone: 44 1 17 929 7481
Fax: 44 1 17 920 0979

techtracking@iop.org

Additional copies of this publication are available from:

Curran Associates, Inc.
57 Morehouse Lane
Red Hook, NY 12571 USA
Phone: 845-758-0400
Fax: 845-758-2633
Email: curran@proceedings.com
Web: www.proceedings.com

TABLE OF CONTENTS

IOP Conf. Series: Materials Science and Engineering 1204 (2021) 011001 doi:10.1088/1757-899X/1204/1/011001

2nd International Symposium on Materials Chemistry (ISyMC'2021)

16 – 20 May 2021

Edited by:

Souad MESDOUR-HAMMADOU
University M'hamed Bougara of Boumerdes, Algeria

Ali-Nordine LEKLOU
University of Nante, France

IOP Conf. Series: Materials Science and Engineering 1204 (2021) 011001 doi:10.1088/1757-899X/1204/1/011001

Preface

The 2nd International Symposium on Materials Chemistry has taken place in Boumerdes Algeria on 16 – 20 May 2021 in the Virtual Scientific Meeting form The Second Symposium on Materials Chemistry (ISyMC2021) aims to bring international and national researchers to high-level exchange on emerging solutions and technologies in the fields of materials science. This international symposium organized by Chemistry department of the University M'hamed Bougara Boumerdes, aims to encourage exchange between participants on the most recent advances in chemical engineering, chemistry materials science, and energetic materials, etc.

Over 951 submissions were submitted to us. These papers have gone through a strict reviewing process performed by our international reviewers. All the submissions were double-blind reviewed, both the reviewers and the authors remaining anonymous. With the hard work from the reviewers, only 230 abstracts were finally accepted. The symposium will cover different topics linked to materials synthesis and characterization, interfacial phenomena, systems and materials optimization and modeling. Leading scientists and researchers will share their recent progress and stimulate discussions on multidisciplinary research activities. These communications were divided into four chapters:

1. Inorganic Materials Chemistry Keywords: concrete and cementitious composites, ceramics matrix and glass, Metals and alloys
2. Organic Materials Chemistry Keywords: polymer matrix composites, nanocomposites, Elastomers and rubber, Paint, varnishes and glues, Resins and solvents...
3. Green Processes Keywords: Wood and paper Ecology, Water treatment, Biomass Valorisation, Sustainable chemistry, Renewable energy, Waste management and recovery, recycling, life cycle analysis and sustainability...
4. Processing and manufacturing technologies Keywords: Experimental techniques standardization, nanotechnologies.

ACKNOWLEDGMENTS

The editors wish to express their gratitude to all the participants of the International Symposium of Materials Chemistry 2021 and its Proceedings: plenary speakers, invited speakers, contributed paper authors and speakers, members of the organizing committee, and members of the scientific committee. Special thanks to all for bringing valuable expertise to pave the way of research into the future.

Dr. IREKTI Amar
ISyMC'2021 Chair
University M'hamed Bougara Boumerdes, Algeria

IOP Conf. Series: Materials Science and Engineering 1204 (2021) 011001 doi:10.1088/1757-899X/1204/1/011001

CONFERENCE CHAIR

Dr. IREKTI Amar, University M'hamed Bougara Boumerdes, Algeria

SCIENTIFIC COMMITTEE

AMRANE Abdeltif, Univ. Rennes 1. France
ABBAS Moussa, Univ. Boumerdes, Algeria
ASMA Farid, Univ. Tizi Ouzou, Algeria
AIT TAHAR Kamel, Univ. Bouira, Algeria
AKNOUCHE Hamid, Univ. Boumerdes, Algeria
AOURAG Hafid, DGRSDT, Algeria
AMRANI Moussa, Univ. Boumerdes, Algeria
AKSIL Tounsia, Univ. Boumerdes, Algeria
BENNICI Simona. IS2 Mulhouse, CNRS. France
BENAMIRA Messaoud. Univ. MSB Jijel, Algeria
BELBACHIR Mohamed, Univ. Oran. Algeria
BOURICHE Rachid, INCC de Bouchaoui. Algeria
BOUTOUCHENT-GUERFI Nabila. CRTSE, Algeria
BOUZID Mohamed, Univ. Boumerdès, Algérie
BOURNANE Mohamed. Univ. Québec Chicoutimi. Canada
BOULOUFA Abdesselam, Univ. Setif 1, Algeria
BUNCIANU Dorel. Timisoara, Roumanie
BRADAI M.Amokran, Univ. Bejaia, Algeria
CHINESTA Francisco. Ecole Centrale de Nantes, France
DJEBARRI Baya, Univ. Boumerdes, Algeria
DJEBRA Nadia, Univ. Boumerdes, Algeria
ELHAFIANE Youssef. ENSCI, Limoges France.
FEDAOUI-AKMOUSSI O. Univ. Tizi Ouzou, Algeria
FERFERA-HARRAR Hafida. Univ. STHBab Ezzouar. Algeria
FRANZEN C. IDK-DENKMAL, Germany
GIACOMO G. Univ. Florence. Italia
GHOUALEM Hafida. Univ. STHBab Ezzouar. Algeria
GUERFI-BOTOUCHENT Nabila, CRTSE, Algeria
GERARD Jean, UR BioWooEB, Montpellier, France
HADDAD Boumediene, Univ. Saïda. Algeria
HADJ-HAMOU Assia Siham. Univ. STHBab Ezzouar.Algeria
HAMMADOU - MESDOUR Souad, Univ. Boumerdes Algeria
IREKTI Amar, Univ. Boumerdes Algeria
JAYAPAL M.R , YIT, CHINA
JOUINI Houda, Univ. Tunis El Manar. Tunisia
JULES T. ASSIH, Univ. Reims Champagne Ardenne-France
KADIRI M. ENSA de Khouribga Marocco
KASIM Rahawi, Univ. of Mosul, Iraq
KADDOUR Omar. FHC, Univ. Boumerdes. Algeria
KHALIL K. Univ. Mohamed I, Maroc
KEBIR Mohammed, CRAPC/(DAES –UR-ADTE) Algeria
KHELOUI-TAOUCHE F., Univ. Tizi Ouzou, Algeria
KHEFFACHE Dj. Univ. Boumerdes, Algeria
KSIBI Zouhaier. Univ. Tunis El Manar. Tunisia
LIMOUSY Lionel, Univ. Haute Alsace Institut Carnot MICA, France
LEBTAHI Hamid, Univ. Ouargla. Algeria
LEKLOU Ali-Nordine, Univ. Nantes. France
LECOMTE Gisele, ENSCI, Limoges France.
LEGENDRE L., Univ. Renne 1. France
MELOUKI Radouane, EMP, Algeria

IOP Conf. Series: Materials Science and Engineering 1204 (2021) 011001 doi:10.1088/1757-899X/1204/1/011001

IOP Conf. Series: Materials Science and Engineering 1204 (2021) 011002 doi:10.1088/1757-899X/1204/1/011002

Peer review declaration

All papers published in this volume of IOP Conference Series: Materials Science and Engineering have been peer reviewed through processes administered by the Editors. Reviews were conducted by expert referees to the professional and scientific standards expected of a proceedings journal published by IOP Publishing.

- **Type of peer review: Single-blind**

Peer review conducted was a single-blind review.

- **Review criteria:**
- **Conference submission management system :** Easychair and
- **Number of submissions received:** 37
- **Number of submissions sent for review:** 37
- **Number of submissions accepted :**11
- **Acceptance Rate (Number of Submissions Accepted / Number of Submissions Received X 100):** 29.72 %
- **Average number of reviews per paper:** 2
- **Total number of reviewers involved :** 18
- **Any additional info on review process :**

Criteria are based on scientific content, clarity and format. Authors must ensure that the papers submitted use the symposium format. There are three review processes: Initial Review, Peer Review and Acceptance Decision.

Initial Review

The Editors-in-Chief evaluates each manuscript in the submission theme to determine if its topic and content can be considered for the conference before being reviewed. Manuscripts that do not meet the initial criteria are rejected.

Peer Review

Papers accepted in the first stage underwent a review and revision process that took an average of two rounds for each paper. The peer review of each paper concentrates on objective and technical concerns to determine whether the research has been sufficiently well conceived, well executed, and well described to justify inclusion in the scientific evidence.After acceptance, articles were reread in English and checked for similarity via Turnitin. Articles with a similarity index greater than 20% were rejected.

Acceptance decision

Based on the reviewers' comments, the editor-in-chief makes a final decision on the acceptability of the manuscript and communicates the decision to the authors, along with the reviewers' reports.

Contact person for queries:

Dr. IREKTI Amar, a.irekti@univ-boumerdes.dz

Antioxidant activity and bioactive compounds of extracts from the Algerian plant *Moltkia ciliata*

Soumaia Chihi[1, 2], Oumelkheir Rahim[1], Ali Douadi[1*] and Rabab Zatout[3]

[1] Pollution & Waste Treatment Laboratory, Kasdi Merbah University, Ouargla 30000, Algeria
[2] VTRS Laboratory, El-Oued University, B.P. 789, El-Oued 39000, Algeria
[3] Department of pharmacy, Faculty of medicine, Benyoucef Benkhedda Algeirs1University, Algiers 16000, Algeria

alidouadi@gmail.com

Abstract. In the present study, we reported the evaluation of antioxidant properties using electrochemical and spectrophotometric assays and determination of total bioactive compounds content of the ethyl acetate (EtOAc) and n-butanol (n-BuOH) extracts of *Moltkia ciliata* growing in Algeria. The results obtained showed that the EtOAc extract was rich in bioactive compounds, which showed the most effective antioxidant capacity in all tests compared to the n-BuOH extract. This was confirmed by the chromatographic analysis and HPLC of the extract of many compounds in different quantities. The difference in antioxidant activity can be explained by differences in the levels of polyphenols and flavonoids. The study concludes this plant is a rich source of phenols and flavonoids, and also showed good in-vitro antioxidant activity by all methods. Thus, the plant *M. ciliata* can be explored as a potential source of natural antioxidant.
Keywords. *Moltkia ciliata*, bioactive compounds, Antioxidant capacity, HPLC, TLC.

1. Introduction

It was recently found that many serious diseases were caused by the use of many synthetic drugs as antioxidants [1]. Several recent studies have proven that the richness of plants in secondary plant metabolites qualifies them to be a source of medicinal agents [2,3]. Phytochemicals have shown several antibacterial, fungal and viral activities [4], as well as their use in several fields such as food, cosmetics, and others [5]. Despite their low molecular weight, polyphenols are responsible for antioxidant properties such as antioxidant, antimicrobial, antiallergic, anti-inflammatory and anticancer [6,7]. Several previous studies reported that several compounds had the property of inhibiting the removal of free radicals, such as phenolic compounds, tannins, anthocyanins and flavonoids, through different mechanisms [8]. *Moltkia ciliata* is a desert plant that is little known in phytochemistry and there is no purely chemical study about it, a contribution from us to shed light on it and to be the first phytochemical study on its effectiveness, this work was devoted to the chemical, antioxidant study of organic extracts of *Moltkia ciliata*, collected using the mixture (methanol/water) applying different analytical methodologies, by answering the following questions: To what extent does this plant contain active substances, especially phenolic compounds? What is the effect of the extraction method on the nature of the extracts? What is the extent of the antioxidant effectiveness of organic extracts?

2. Materials and methods

2.1. Chemicals and reagents
All chemicals, solvents and reagents were analytical grades or purest quality purchased and all used as received.

2.2. Plant Material

Moltkia ciliata is a species of plants in the family Boraginaceae. This family consists of about 131 genera, most of which are annual, semi-annual or perennial herbs, sub-shrubs, and the plants of this family are widespread in tropical and subtropical regions [9]. *M. ciliata* is a small, long-stemmed shrub that is wrapped with hard bristles, no longer than 30 cm long [10]. This plant is a food sanctuary for some desert animals especially camel; it has several traditional uses in the medical field used for abdominal diseases, wound healing and treatment against scorpion stings *M. ciliata* has attracted the interest of many researchers, especially in the statistical studies of some characteristics of desert plants [11].

Figure 1. Outside of the plant *Moltkia ciliata*

2.3. The phytochemical study
2.3.1. Phytochemical Screening
We performed several preliminary phytochemical tests to determine the different active substances present in the aerial part of the plant [12-14].

2.3.2. Extraction and isolation

The aerial parts of the plant (100 g) were soaked in the mixture (MeOH/H_2O) (80/20, v/v) (99 °C; Biochem Chemopharma (Canada)) for 24 h, the process was repeated three times, the filtrate was collected, and concentrated at 35 °C and then dissolved in distilled water. The resulting solution was treated by a series of successive liquid-liquid extractions using solvents of increasing polarity, starting with petroleum ether, then chloroform, then ethyl acetate and finally with n-butanol, phases are dried Organic magnesium sulphate anhydrous (97%, Acros Organics), and then its concentration.

2.3.3. Chromatographic Separation
Thin layer chromatography (TLC) is a quick and simple analytical technique used for the purpose of separating and identifying metabolites based on adsorption phenomena. This technique provides an overview of the compounds present in the extract [15].

2.3.4. Phenolic identification
By high-performance liquid chromatography HPLC (Shimadzu, LC 20, Japan) [16], the mobile phases were: (A: acetic acid, B: acetonitrile (v/v)), at rate (0.8 ml/min) and the detection wavelength was 285nm. Phenolic compound standards querecetin, vanillin, gallic acid, caffeic acid, ascorbic acid, rutin and chlorogenic acid were dissolved in solvents extraction and used for identification of the phenolic

acids present in different extracts of *M. ciliata*. Using the peak area of the reference compounds the quantification of individual phenolic compounds in the extracts is performed and expressed in mg/g of the extract.

2.4. Determination of total bioactive compounds content

2.4.1. Total phenolic content

The folin-ciocalteu reagent (FCR; 98%; biochem chemopharma (Canada)) was used to determine the total concentrations of the phenolic content of the extracts by following the method described by (Singleton Rossi 1999) [13], where we use the gallic acid ($C_6H_2(OH)_3COOH$; 99%; Alfa Aesar (Germany)) as reference phenol. 1ml of extract solution add 0.5ml of dilute folin ciocalteu reagent 10 times, after 5 min add 2 ml of sodium carbonate solution (Na_2CO_3, 7.5%) the mixture completed the reaction for 30 min at room temperature. The absorbance was read at 765 nm using a UV-Vis spectrophotometer (Shimadzu Scientific Instruments, Kyoto, Japan). Using the standard compound curve,the total phenol content was calculated and expressed in mg of gallic acid equivalent per gram of dry weight of the extract [8,17].

2.4.2. Total flavonoids content

The total flavonoids content were determined using the method of (Ordonez et al 2006) [18], where we use the rutin ($C_{27}H_{30}O_{16}$; 97 %; Alfa Aesar (Germany)) as reference flavonoid. A volume of 0.5 ml of 2% $AlCl_3$ methanol solution was added to 0.5 ml of extract solution. The mixture was incubated for 1 h at room temperature for yellow color appearance; the absorbance was measured at 420 nm. The total flavonoids content was calculated from the calibration curve, and the results were expressed as mg of routine equivalent per g dry weight of extract [19].

2.4.3. Total flavanol content

The total flavanols content were determined using the method of Kumaran and Karunakaran (2007) [20], where we use the quercetin ($C_{15}H_{10}O_7$; 97%; Alfa Aesar (Germany)) as reference flavanol. 2.0 ml of the sample was mixed with 2.0 ml of 2% $AlCl_3$ prepared in methanol and 3.0 ml of 50 g/l sodium acetate solution were added. The mixture was incubated at 20 °C for 2.5 h after which the absorption was read at 440 nm using spectrophotometer. The total flavanol content was calculated from the calibration curve, and the results were expressed as mg of quercetin equivalent per g dry weight of extract [21]. The total experiment was conducted in triplicate and the results were expressed as means ± SD (standard deviation).

2.5. Antioxidant activities

2.5.1. Free radical scavenging activity

2.5.1.1. Free 2,2-diphenyl-1- picrylhydrazyl (DPPH) radical-scavenging ability

Free radical scavenging activity of extracts were measured using a stable 2,2-diphenyl-1-picrylhydrazyl radical (DPPH) ($C_{18}H_{12}N_5O_6$;97(%); Sigma-Aldrich Chemicals (St Louis, MO, USA)) and was determined spectrophotometrically by the slightly modified method of Djeridane et al. 2010 [22]. This test is done by mixing 1.5 ml of 100 μM DPPH in methanol with 1.5 ml of different concentrations of a solution of extracts and standards in the same solvent. Incubate in the dark for 30 minutes at room temperature. The absorbance is read in a UV-Vis spectrophotometer at 517 nm.

Using the following equation (1), we were calculated the ability of the test sample to quench the root of DPPH.

$$\textbf{DPPH scavenging activity } (\%) = \frac{\textbf{Absorbance of control} - \textbf{Absorbance of sample}}{\textbf{Absorbance of control}} \times 100 \quad (1)$$

The activity strength of the samples was expressed as the IC50 concentration, which is the amount of antioxidant needed to inhibit 50% of the initial amount of DPPH [23].

2.5.1.2.Free superoxide anion radical ($O_2^{\cdot-}$) radical-scavenging ability

By using the electrochemical method that depends on the techniques of measuring periodic voltmeters, the antioxidant capacity of the studied extracts was measured, using PGP301 potentiostat (radiometer analytical SAS) connected to an electrochemical cell of three electrode [24,25]. The antioxidant capacity is calculated by the current density of the anodic curve of a voltammogram [25,26]. The superoxide anion radical was generated by one electron reduction of the commercial molecular oxygen (O_2) dissolved in DMF (analytical grade, Sigma-Aldrich) containing 0.1 M TNBHFP at room temperature (28±1)°C, were used quercetin as a standard in the calculation of antioxidant capacity of the studied samples. The ability of the test sample to quench superoxide anion radicals (O_2^{-*}) (% Inhibition of (O_2^-)) was calculated using the equation (2) [27,28], and according to the researcher's method Bourvellec et al. [29]

$$O_2^- \text{radical scavenging activity } \% = \frac{i_0 - i_s}{i_0} \times 100 \quad (2)$$

Where i_0 and i_s are the anodic peak current densities of the superoxide anion radical in the absence and in the presence of the extract.

2.5.2. The ferric reducing antioxidant power (FRAP) assay

FRAP assay was performed according to the method of Benzie and Strain (1996) with slightly modification [30]. The stock solutions included 300 mM acetate buffer (3.1 g $C_2H_3NaO_2$-$3H_2O$ and 16ml $C_2H_4O_2$), pH 3.6, 10mM TPTZ (2, 4, 6-tripyridyl-s-triazine; $C_{18}H_{12}N_6$;98(%); Sigma-Aldrich Chemicals (St Louis, MO, USA)) solution in 40 mM HCl, and 20 Mm $FeCl_3.6H_2O$ solution. The FRAP solution was prepared by mixing 25ml acetate buffer, 2.5ml TPTZ solution, and 2.5ml $FeCl_3.6H_2O$. The ascorbic acid ($C_6H_8O_6$; (99%); Alfa Aesar (Germany)) used a standard compound. 0.3 ml of ascorbic acid solutions was mixed with 0.27 ml of the FRAP solution. The mixture was Leave in the dark for 30 min and gets a violet color. The absorbance was measured at 593 nm in a spectrophotometer. All determinations were carried out in triplicate. Using the standard curve relationship of ascorbic acid, we evaluated the antioxidant activity of organic extracts and the equivalent amount of 1 g of the extract, this was calculated the following equation (3) [31]:

$$\text{FRAP value} = \frac{A_{593nm} \text{ of test sample(conc. 1000 } (\mu g/ml)) \times FeCl_3 \text{concetration}(\mu mol/l)}{4min\ A_{593nm} \text{ of } FeCl_3 \text{(at conc. } 1000(\mu g/ml))} \quad (3)$$

2.5.3. Total antioxidant capacity (Molybdate ion reduction)

The total antioxidant capacity of the extracts was evaluated by phosphomolybdenum method [32]. In this method, the antioxidant compounds convert Mo (VI) to Mo (V), it is green in color [33]. The ascorbic acid used a standard compound. The reagent solution was prepared by mixing (sulphuric acid 0.6 M, sodium phosphate 28 mM (Rathburn Chemical (Walkerburn, Peebleshire, UK)), ammonium molybdate 4 mM ($(NH_4)_2MoO_4$; (98%); BioChem (Quebec, Canada)). 0.3 ml of ascorbic acid solutions was mixed with 3 ml of the detergent solution, leave it for an hour in a water bath temperature of 95 °C, we get the green color [34]. Using the standard curve relationship of ascorbic acid, we evaluated the antioxidant activity of organic extracts and the results were expressed as ascorbic acid equivalent (AAE) in milligrams per gram dry extract.

2.5.4. ABTS Radical Cation Decolorization Assay

The cation radical scavenging activities of the extracts were determined using ABTS 2,2′-azino-bis-(3-ethylbenzothiazoline-6-sulphonic acid) diammonium salt ($C_{18}H_{24}N_6O_6S_4$;98(%); Sigma-Aldrich Chemicals (St Louis, MO, USA)) (Pellegrini and al., 1999) [35]. This method is based on removing the greenish blue color of the cationic radical in the presence of an antioxidant molecule at a wavelength of 734 nm [36]. For ABTS assay, the stock solutions included 7 mM ABTS + solution and 2.4 mM potassium persulphate solution. The working solution was then prepared by mixing the two stock solutions in equal quantities and allowing them to react for 12 h at room temperature in the dark. The solution was then diluted by mixing 1 ml ABTS + solution with 60 ml methanol to obtain an absorbance of (0.706 ± 0.001) units at 734 nm using the spectrophotometer. Fresh ABTS + solution was then prepared for each assay. Plant extracts/standard (BHT; butylated hydroxytoluene; 99(%); Sigma-Aldrich Chemicals (St Louis, MO, USA)) (1 ml) of different concentration was allowed to react with 1 ml of the ABTS+ solution freshly prepared solution and the absorbance was taken at 734 nm after 7 min using the spectrophotometer [34]. The ABTS scavenging capacity of the extracts was compared with that of standard and percentage inhibition calculated using the following formula [37].

$$\text{ABTS radical scavenging activity } (\%) = \frac{Abs_{control} - Abs_{sample}}{Abs_{control}} \times 100 \quad (4)$$

IOP Conf. Series: Materials Science and Engineering 1204 (2021) 012001 doi:10.1088/1757-899X/1204/1/012001

$Abs_{control}$: The absorbance of ABTS radical + methanol. Abs_{sample}: The absorbance of ABTS radical + extracts/standard.

2.6. Statistical Analysis
Experimental values are given as means ± standard deviation (SD) of three replicates for antioxidant activity. Statistical calculations were carried out by OriginPro Version 9.1 software (OriginLab Corporation). Values of $p < 0.05$ were regarded as significant.

3. Results and discussion

3.1. Phytochemical study
3.1.1. Preliminary phytochemical
The obtained results (Table 1) showed the richness of this plant in most of the natural compounds, previous studies of plants of this genus (*Moltkia*) have proven that they contain most of these effective compounds

Table 1. Phytochemical constituents of *M. ciliate*

Compounds	Results
Flavonoids	(+++) Appearance of a red color characteristic aglycone flavones
Alkaloids	(++) appearance of turbidity
Sterols	color changes from purple to blue or green(++)
Volatile oils	(+) A distinctive odor emission
Saponosides	(++) Appearance of foam shaken in for15 min
Tannins	(+++) Appearance of a blue-green color
Coumarins	(-) no precipitate
Cardenolides	(-) no precipitate

(+++): important Presence, (++): average Presence, (+): weak, presence and (-): absence

3.1.2.Extraction yield
It is calculated by the following equation (5),

$$R\% = (m_f / m_i) \times 100 \quad (5)$$

Where: m_f; The mass of the organic extract obtained and m_i; The mass of the plant sample

Note the difference in extraction yield between organic extracts; the polarity of both the solvent and the component affects the separation [38]. The extracts AcOEt and *n*-BuOH are of the highest value, where the study was confined around them.

3.1.3. HPLC analysis

The constituents in the different extracts were analyses by HPLC. The contents of these components in different extracts were determined according to the calibration curves, where y was the peak area and x was the concentration of analyte (0–80 µg/ml). The quantitative results are summarized in Figure 2, so that represents: A.A: Ascorbic Acid; G.A: Gallic Acid; Ch.A: Chlorogenic Acid; C.A: Caffeic Acid; V: Vanillin; Q: Quercetin; R: Rutin. As shown, all compounds are present in the extracts with different quantities. Rutin, caffeic acid, quercetin and chlorogenic acid are present in significant quantities. Ethyl acetate extract is richer in compounds than others; this is due to the difference of these compounds in terms of polarity.

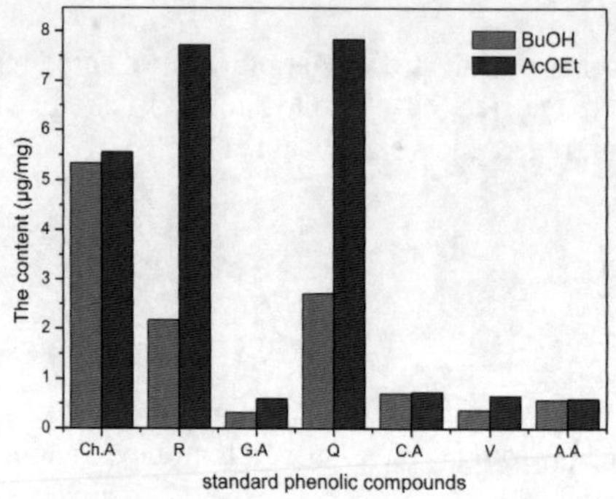

Figure 2. Constituents content analyses by HPLC

3.1.4. Chromatographic separation

Chemical constituents of the extracts were separated on aluminum-backed thin layer chromatography (TLC) plates (Merck, silica gel 60, F254).the solvent system were used: (methanol/chloroform/eau; 5/20/0.5), (methanol/chloroform; 1/3), (methanol/chloroform/n-butanol; 1/1/1). The separate chemical compounds were examined using ultraviolet rays 254 nm and 365 nm and chemical reagents, and it was possible to predict the possibility of the presence of the following compounds [39,40]:

Flavon or flavonol containing OH at position C_5 and OH at position C_4 replaced or deleted purple, some flavonoids contain OH in position C_5 or chalcones containing OH in position C_4 and lack OH on the aromatic ring B(brown), isoflavones do not contain OH at position C_5 free radiant blue, flavonol containing OH at position C_3 with or no free OH at position C_5 Yellow. Where 9 compounds were identified in (*n*-BuOH) extract, 13 compounds in (AcOEt) extract.

3.2. Total bioactive compounds content

Figure 3 represent the analytical data for phenolics content, flavonoids and flavanol of the extracts of *M. ciliata*.In comparison, AcOEt extract showed the higher quantities compared to n-BuOH extract. The AcOEt extract contains a large number of polyphenolic compounds and flavonoids mainly due to the polarity of polyphenols determined [41,42]. These compounds can have some pharmacological effects as well as beneficial for human consumption [43,44].

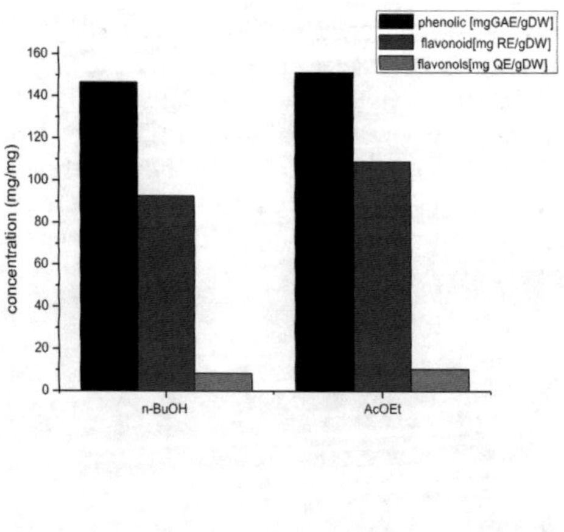

Figure 3. Total phenolic content, flavonoid and flavonols of extracts from *M. ciliate*

3.3.Antioxidant activity

Various methods were used to measure the antioxidant capacity of the studied extracts and the results obtained are shown in Table 2.

Table 2. Antioxidant activity of *M. ciliata* extracts as determined by the DPPH, FRAP, TAC, superoxide anion($O_2^{\cdot-}$), ABTS assays

Extracts	IC_{50}DPPH (mg/ml)	FRAP Value (mg /g)	IC_{50} superoxide anion($O_2^{\cdot-}$) (mg/ml)	IC_{50}ABTS (mg/ml)	TAC Value (mg /g)
Standard	**0.28 ± 0.03**	-	**0.23±0.01**	**0.093±0.02**	-
AcOEt	**0.06 ± 0.03**	292.29 ± 0.02	**0.51±0.01**	**0.09±0.01**	**91.03 ± 0.04**
***n*-BuOH**	**0.10 ± 0.01**	**256.68 ± 0.02**	**1.02±0.03**	**0.23±0.01**	**85.61 ± 0.05**

3.3.1. Free radical scavenging activities study

3.3.1.a. Free 2,2-diphenyl-1-picrylhydrazyl (DPPH) radical-scavenging ability

The DPPH radical scavenging ability of plant extracts and ascorbic acid standard shown in Figure 4, we note that increasing the concentration leads to an increase in inhibition, it is clear from the results obtained that the AcOEt extract shows the highest DPPH radical scavenging activity (0.20±0.02) which was significantly closer to that of standard antioxidant. The radical scavenging ability of the extract is related to the presence of total phenolic content (TPC), because these compounds contribute significantly to the antioxidant activity and act as free radical scavengers due to their redox properties [45].

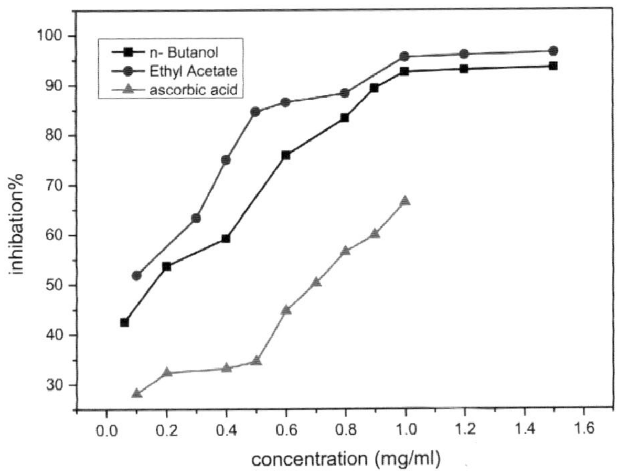

Figure 4. The DPPH radical scavenging activity of extracts and the ascorbic acid

3.3.1.b. Free superoxide anion radical ($O_2^{\cdot-}$) radical-scavenging ability

The following procedure was followed for the measurement of the antioxidant capacity: It was obtained cyclic voltammograms of a solution of (DMF + 0.1M TNBHFP) recorded at a scan rate of 100 mV/s on GC at 28°C (A) saturated with commercial oxygen and are shown in Figure 5, and the same way this has been done cyclic voltammograms of extracts and the standard are shown in Figure 6, then we calculate the percentage of inhibition.

IOP Conf. Series: Materials Science and Engineering 1204 (2021) 012001 doi:10.1088/1757-899X/1204/1/012001

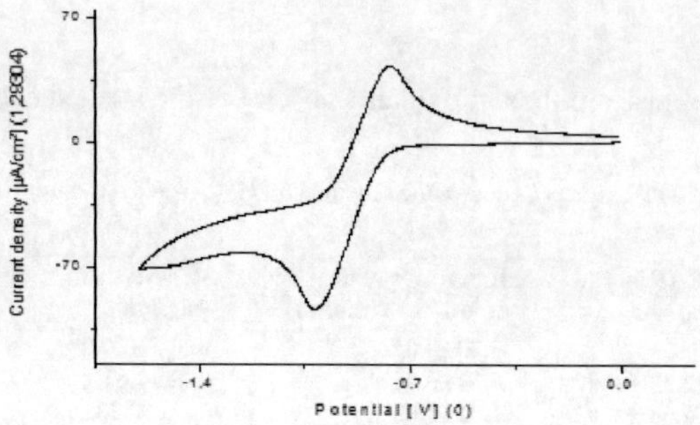

Figure 5. Cyclic voltammograms of a solution of (DMF + 0.1 M TNBHFP) recorded at a scan rate of 100 mV/s on GC at 28°C (A) saturated with commercial oxygen

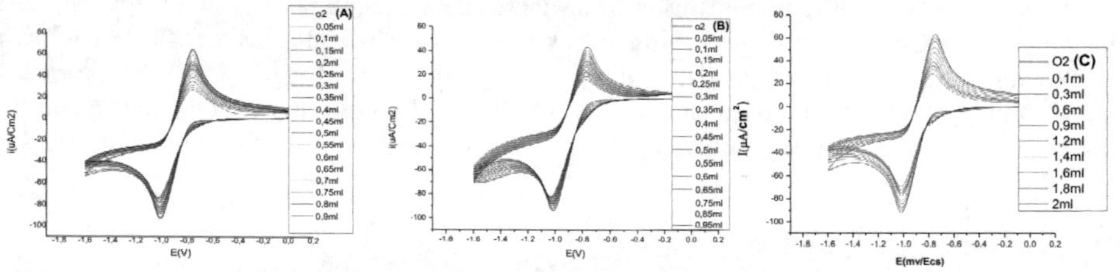

Figure 6. Cyclic voltammograms of O_2^{-} in presence of (A)AcOEt extract,(B) n-BuOH extract and (C) Quercetin at different concentrations in DMF +0.1M TNBHFP on GC as working electrode vs. Hg/Hg2Cl as reference at 28 °C with scan rate of 100 mV/s

The cyclic voltammograms shown in Figure 6 show that the addition of *M. ciliata* extracts always causes a decrease in the current density of the anode peaks ip_a accompanied by a shift in the potential of the anode peaks to the more positive values. It is known that the addition of the extract containing flavonoids to the solution containing the superoxide radicals leads to a decrease in the current (at least in one stream; cathodic or anodic) and a positive shift in the peak potential value, and this is due to the scavenging activity ofthe added flavonoids and the studied extracts are rich in them [26]. The concentration of the radical is directly proportional to the decrease in the current and is a direct measure of the antioxidant activity [46].

3.3.2. The ferric reducing antioxidant power (FRAP) assay

EtOAc extract gave the best effective compared to *n*-BuOH extract, and this is because the flavonoids and phenolic acids present in the extracts exhibit strong antioxidant activity which is manifested in their ability to compound with mineral atoms, especially iron and copper [47]. The antioxidant compound present in the samples forms a colored complex with potassium ferricyanide, trichloroacetic acid and ferric chloride, which is measured at 700 nm by UV-Spectrophotometer [48].

3.3.3. Total antioxidant capacity (Molybdate ion reduction assay)

The total antioxidant capacity of studied M. ciliata extracts was determined by molybdate ion reduction method The high antioxidant activity of these plant extracts is due to their flavonoid and phenol content [49].

3.3.4. ABTS radical scavenging activity

Bleaching of ABTS cation has been extensively used to evaluate the antioxidant capacity [45]. The extracts from *M. ciliata* were fast and effective scavengers of $ABTS^{\cdot+}$ radical Figure 7 and the obtained results are shown in Table 2. This activity is comparable with that of BHT, the standard antioxidant used in this study. The percentage inhibition was 78.8%, 96.7% and 80.2% in n-butanol, ethyl acetate and BHT respectively at 0.5 mg/ml, the highest concentrations tested. Higher concentrations of the extracts were more effective in quenching free radicals in the system.

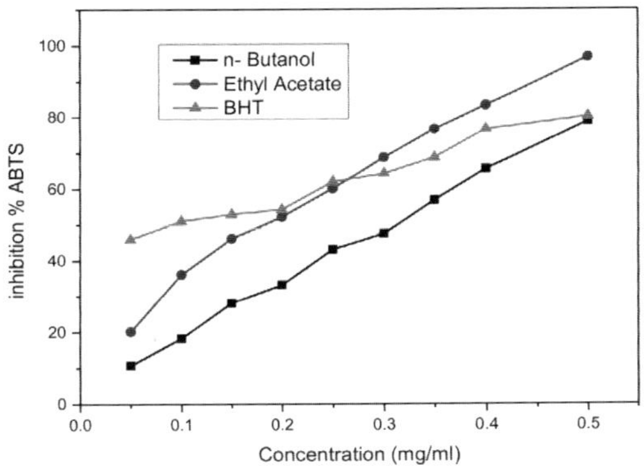

Figure 7. ABTS radical scavenging activities of extracts of *M. ciliata*

EtOAc extract gave the best effective compared to *n*-BuOH extract. It turns out that there is a relationship between $ABTS^{\cdot+}$ radical scavenging activity for extracts and Total phenolic, flavonoids content and confirms this compounds have contributes directly to the antioxidant effects of these extracts.

4. Conclusion

The ethyl acetate extract exhibited the highest activity in the all assays followed by *n*-butanol extract.This also proves that polyphenolic compounds and antioxidant activities are closely linked. The results obtained herein are in agreement to a certain degree with the traditional uses of *M. ciliata*as a valuable source for antioxidant drugs.

Acknowledgements

The authors would like to thank to Algerian Ministry of Higher Education and Scientific Research and General Directorate of Scientific Research and Technological Development (DGRSDT) for their support and providing the necessary facilities to carry out this research.

References

[1] Tzima K, Brunton N P and Rai D K 2018 *Plants* **7(2)** 1.
[2] Klejdus B and Kováčik J 2016 *Industrial Crops and Products* **83** 774.
[3] Leja K B and Czaczyk K 2016 *Acta Scientiarum Polonorum Technologia Alimentaria* **15(4)** 353.
[4] Gopi Krishnan S, Ajith A P, Renish C, Duraiselvan S and Subash Maruthu Pandian K 2018 *Int. J. S. Res. Sci. Engg. Tech.* **4(8)** 443.
[5] Engel R, Szabo K, Abranko L, Rendes K, Fuzy A and Takas T 2016 *Journal of agricultural and food chemistry* **64(19)** 3733.
[6] Quideau S, Deffieux D, Douat-Casassus C and Pouysegu L 2011 *Angewandte Chemie International Edition* **50(3)** 586.
[7] Oliveira A S, Ribeiro-Santos R, Ramos F, Castilho M C and Sanches-Silva A 2018 *Food analytical methods* **11(2)** 440.
[8] Singh S and Vyas M 2016 *International Journal of Pharmaceutical Sciences and Research* **7(4)** 1793.
[9] RaNjbaR M and Khalvat I S 2019 *Phytotaxa* **408(3)** 143.

[10] Odeja O, Ene Ogwuche C, Emeka Elemike E and Obi G 2017 *Clinical Phytoscience* **2(1)** 1.
[11] Cohen J I 2014 *Cladistics* **30(2)** 139.
[12] Savithramma N, Rao M L and Ankanna S 2011 *International Journal of Research in Pharmaceutical Sciences* **2(4)** 643.
[13] Singleton V L, Orthofer R and Lamuela-Raventós R M 1999 *Methods in enzymology* **299** 152.
[14] Shikha R and Savaita D 2017 *International Journal of Natural Products Research* **7(2)** 7.
[15] Hadj Salem J 2009 *Doct. Thesis* (France: Institut National Polytechnique de Lorraine) p 251
[16] Özen T 2010 *Grasas y aceites* **61(1)** 86.
[17] Kivrak Ş, Göktürk T, Kivrak I, Kaya E and Karababa E 2019 *Food Science and Technology* **39(2)** 423.
[18] Ordonez A, Gomez J and Vattuone M 2006 *Food chemistry* **97(3)** 452.
[19] Phull A R, Abbas Q, Ali A, Raza H, Zia M and Haq I U 2016 *Future Journal of Pharmaceutical Sciences* **2(1)** 31.
[20] Kumaran A and Karunakaran R J 2007 *LWT-Food Science and Technology* **40(2)** 344.
[21] Mbaebie B, Edeoga H and Afolayan A 2012 *Asian Pacific Journal of Tropical Biomedicine* **2(2)** 118.
[22] Djeridane A, Yousfi M, Brunel J M and Stocker P 2010 *Food and Chemical Toxicology* **48(10)** 2599.
[23] Santos C M and Silva A 2020 *Molecules* **25(3)** 696.
[24] Ghiaba Z, Yousfi M, Hadjadj M, Saidi M and Dakmouche M 2014 *International Journal of Electrochemical Science* **9(2)** 909.
[25] Anjomshoa M, Fatemi S J, Torkzadeh-Mahani M and Hadadzadeh H 2014 *Spectrochimica Acta Part A: Molecular and Biomolecular Spectroscopy* **127** 511.
[26] Douadi K, Chafaa S, Douadi T, Al-Noaimi M and Kaabi I 2020 *Journal of Molecular Structure* **121 7**128305.
[27] Wu J, Bi S Y, Sun X Y, Zhao R, Wang J H and Zhou H F 2019 *Journal of Biomolecular Structure and Dynamics* **37(13)** 3496.
[28] Hui N, Niu X L, Han J Y, Sun W and Jiao K 2010 *Amino acids* **38(3)** 711.
[29] Le Bourvellec C, Hauchard D, Darchen A, Burgot J L and Abasq M L 2008 *Talanta* **75(4)** 1098.
[30] Benzie I F and Strain J J 1996 *Analytical biochemistry* **239(1)** 70.
[31] Franco D, Munekata P E, Agregán R, Bermúdez R, López-Pedrouso M, Pateiro M and Lorenzo J M 2020 *Antioxidants* **9(2)** 90.
[32] Celep E, Aydın A, Kırmızıbekmez H and Yesilada E 2013 *Food and chemical toxicology* **62** 448.
[33] Rubio C P, Hernández-Ruiz J, Martinez-Subiela S, Tvarijonaviciute A and Ceron J J 2016 *BMC veterinary research* **12(1)** 1.
[34] Kamath S D, Arunkumar D, Avinash N G and Samshuddin S 2015 *Advances in Applied Science Research* **6(6)** 99.
[35] Re R, Pellegrini N, Proteggente A, Pannala A, Yang M and Rice-Evans C 1999 *Free radical biology and medicine* **26(9-10)** 1231.
[36] Coulombier N, Nicolau E, Le Déan L, Antheaume C, Jauffrais T and Lebouvier N 2020 *Marine drugs* **18(2)** 122.
[37] Youn J S, Kim Y J, Na H J, Jung H R, Song C K, Kang S Y and Kim J Y 2019 *Food science and biotechnology* **28(1)** 201.
[38] Kobus-Cisowska J, Szczepaniak O, Szymanowska-Powałowska D, Piechocka J, Szulc P and Dziedziński M 2020 *Ciência Rural* **50(1)**.
[39] Rahim O, Hamra H, Guerrah S and Allaoui M 2015 *Revue des Bioressources* **5(1)** 188.
[40] N'gaman K C C, Békro Y A, Mamyrbékova-Békro J A, Bénié A and Gooré S 2009 *Eur. J. Sci. Res.* **36(2)** 161.
[41] Felhi S, Saoudi M, Daoud A, Hajlaoui H, Ncir M, Chaabane R, El-Feki A F, Gharsallah N and Kadri A 2017 *Food Science and Technology* **37(4)** 558.
[42] Tian S, Shi Y, Zhou X, Ge L and Upur H 2011 *Pharmacognosy magazine* **7(25)** 65.
[43] Bakari S, Hajlaoui H, Daoud A, Mighri H, Ross-Garcia J M, Gharsallah N and Kadri A 2018 *Food Science and Technology* **38(2)** 310.
[44] Zeng L B, Zhang Z R, Luo Z H and Zhu J X 2011 *Food chemistry* **125(2)** 456.

[45] Tayag C M, Lin Y C, Li C C, Liou C H and Chen J C 2010 *Fish & shellfish immunology* **28(5-6)** 764.
[46] Ahmed S and Shakeel F 2012 *Czech Journal of Food Sciences* **30(2)** 153.
[47] Rabeta M and Faraniza R N 2013 *International Food Research Journal* **20(4)** 1691.
[48] Phatak R S and Hendre A S 2014 *Journal of Pharmacognosy and Phytochemistry* **2(5)** 32.
[49] Khiya Z, Oualcadi Y, Gamar A, Amalich S, Berrekhis F, Zair T and Hilali F E 2019 *Phytothérapie* **17(6)** 321.

Valorization of potato peel residues to produce a bioflocculant to be used in the treatment of liquid effluents

Fatima Zohra CHOUMANE [1, 2]*, Fatiha ZAOUI[3], Fatma KANDOUCI [1], Bouhana MAACHOU[1] and Belkacem BENGUELLA [2]

[1] Department of Process engineering, Faculty of Technology. University of Saida- Dr Moulay Tahar, Algeria

[2] Laboratory of Inorganic Chemistry and Environment, Department of Chemistry, Faculty of Sciences,University of Tlemcen - Abou Bakr Belkaid, , Algeria

[3] Department of chemistry, Faculty of science. University of Saida- Dr Moulay Tahar, Algeria

fatimazohra.choumane@univ-saida.dz

Abstract. The present study aims primarily to investigate the flocculation capacity of a novel potato peel-based bioflocculant in wastewater treatment. The analysis of wastewater revealed high COD and BOD_5 contents that could respectively reach the values 529.08 mg O_2/l and 317.03 mg O_2/l. In addition, the effect of experimental parameters such as the pH, coagulant/flocculant dosage, and contact time, was studied using the coagulation-flocculation treatment technique. The *experiments were carried out with* a lab-scale *jar-test apparatus* where aluminum chloride ($AlCl_3$) was used as a coagulant agent. It should be noted that the optimal dose of $AlCl_3$ was equal to 0.6 g at 10 mn, with turbidity reduction of 99.01%. Moreover, the flocculation tests showed higher turbidity removal (98%), for a flocculant dosage of 0.2g. The results obtained in this study indicated that using the natural potato peel-based flocculant, in the presence of a coagulant, provides an effective and environmentally friendly coagulation option for wastewater treatment.

Keywords: coagulation-flocculation, wastewater, turbidity, treatment

1. Introduction

No one denies that industrial sites throughout the world remain the primary cause of water pollution due to the diversity of industries (textiles, paints, pharmaceuticals, metal coatings, corrosion inhibition, etc.) [1-5]. Nowadays, various wastewater treatment technologies, such as adsorption, oxidation, chemical precipitation, etc., with different degrees of effectiveness and success, are available to control and minimize water pollution [6, 7]. Each one of these techniques, when applied, showed some advantages and limitations [8]. However, many of these methods have not been widely used due to high costs, formation of hazardous by-products and intensive energy requirements [9]. It is worth noting that the coagulation-flocculation process is being widely applied in raw water treatment, and in particular wastewater treatment. This technique is broadly used because it relatively low-cost, robust and environmentally friendly [10].

The coagulation-flocculation technique has been widely used in multiple industry sectors so far. This process is generally applied for the removal of colloidal particles, soluble compounds, and vey fine solids suspended in solutions by colloid destabilization which leads to the formation of larger particles that are commonly referred to as flocs [11, 12]. Currently, this technique can be effectively achieved through the use of inorganic coagulants and fossil-based organic flocculants. Note that flocculants are particularly advantageous because they are efficient at low dosages and able to form quite large flocs [13, 14].

However, it was shown that the use of synthetic flocculants causes serious environmental damage and health problems. In addition, this method generates a great deal of controversy related to the production of large volumes of toxic sludge and the dispersion of acrylamide oligomers, which engenders serious health hazard due to its harmful carcinogenic and neurotoxic effects on human beings [15-17]. For these reasons, as an alternative, bioflocculants have attracted increasing attention for water treatment due to their biodegradability, non-toxic properties, and remarkable flocculation performances that are sometimes comparable to those of synthetic flocculants [18-22]. This eco-friendly approach involves biodegradability and sustainability with the exploitation and utilization of byproducts derived from human activities such as fishing, agriculture, industry, and others. It is worth citing some examples of bioflocculants like alginate, cellulosic materials, starch, chitosan, xanthan gum, Moringa oleifera, okra, guar gum and Cassia tora gum, and tannins, which represent economical, viable, and safe substitutes for synthetic flocculants [15-17]. It is widely admitted that potato peel waste causes much environmental pollution when stored outside without any appropriate and safe disposal. In addition, potato peel processing can release potential pollutants that cause serious environmental problems. Consequently, due to the multifunctional nature of potato peels, it is highly urgent and essential to design and develop environmentally friendly processing procedures to deal with these wastes [23]

On the other hand, it is important to indicate that some food byproducts, like potato peels, possess essential organic matter. In this context, Gebrechristos et al. (2018) [23] indicated that potato peel residues can potentially be used as a food preservative, pharmaceutical ingredient, animal food for the purpose of promoting eco-friendly food industries; they also reported that potato peels can also employed in renewable energy. Similarly, the author [24] found out that potato peel waste is an interesting low cost adsorbent which is quite efficient in the adsorption of the methylene blue (MB) dye from aqueous solutions. Indeed, the results obtained showed that the percentage of elimination of this dye can go from 61.4 to 87.6% while increasing the adsorbent mass. Based on these findings, it can be said that potato peels could be effectively used as an eco-friendly adsorbent material for the removal of methylene blue dye from an aqueous solution. Therefore, potato peel residues can be valorized to produce a flocculant agent for the treatment of wastewater through the coagulation-flocculation procedure.

The primary purpose of this paper is to investigate the effect of some operational factors on the treatment of water using potato peel waste as a bioflocculant. It is worth indicating that, over the past few years, biopolymer-based flocculants have been attracting *increasing interest* from many researchers because they are easily biodegradable and are highly environmental friendly. However, natural flocculants, which are characterized by a moderate flocculating efficiency and short shelf life, should be used at large doses [13]. Moreover, factors such as the pH, coagulant dose and coagulant type, play an essential role in the coagulation and flocculation operations. These factors were investigated in order to determine the optimal efficiency under the best experimental conditions and to identify the most appropriate mechanisms that could be involved in the treatment process. These flocculants turned out to be highly efficient in removing COD, turbidity, color and metals, depending on the types of contaminant and coagulant/flocculant used [25].

2. Materials and Methods

2.1. Preparation of potato Peel

Nowadays, food processing industries are producing huge amounts of waste. In particular, potato processing factories are rapidly expanding and are generating enormous volumes of potato peel waste. When decomposed, this potato byproduct can cause serious environmental pollution **[23].** For this reason, potato peel has been investigated by many researchers as a good source for biopolymer production. In addition, it was found that potato peel waste represents an interesting source of vegetable starch. It should also be noted that this type of starch swells well and possesses high viscosity.

For the purpose of understanding the physicochemical properties of potato peels, it is required to determine their physical and chemical composition.

It was revealed that potato peels contain various polyphenols and phenolic acids, which are responsible for its antioxidant activities, as well as fatty acids and lipids, which exhibit interesting antibacterial activities [26, 27]. In addition, potato peels contain starch (25%), non-starch polysaccharides (30%), proteins (18%), acid-soluble and acid-insoluble lignin (20%), lipids (1%) and ash (6%) on dry matter basis [28]. Furthermore, the lipid fraction includes long-chain fatty acids, alcohols, triglycerides and sterol esters. Moreover, lignin units were also found in the cell wall of potato [29]. It was also found that potato peel waste is rich in starch (52% dry weight), but the content of fermentable reducing sugar is quite low (0.6% dry weight) [30].

In this study, potato peel is used as a bioflocculant without any modification or prior purification. For this, the potato peels to be tested were first washed with tap water and then dried in the open air for several days. Afterwards, they were cut into small pieces and finely crushed to powder. Then, the samples thus obtained were placed in bottles and stored at room temperature.

3. Experimental

3.1. Characterization

In order to characterize the potato peels, to be used as a bioflocculant, Fourier-transform infrared spectroscopy analyses were carried out to study the surface functional groups of this material using potassium bromide pellets. The infrared analysis was performed with a PerkinElmer Spectrum One FT-IR Spectrometer version 10.4.1.

3.2. Wastewater

The wastewater used in this study was brought from the wastewater treatment plant in the small town of Rebahia, located near wadi Saida, in the northwestern part of the City of Saida. The procedures applied for the physicochemical measurements were carried out in accordance with the Standards described by the French national organization for standardization (AFNOR).

The turbidity is expressed in nephelometric turbidity units. In addition, factors such as the pH, conductivity and turbidity were measured using a Hanna Instruments pH meter, a conductivity meter and an infrared AQUA LYTIC Model AL250T-IR turbidity meter, respectively. The results of the physicochemical measurements of the wastewater under study are summarized in Table 1.

Table1. Results of the characteristics of wastewater analyzed

Parameter	Units	Average values	AFNOR standards	Algerian standards
pH	-	8.12	NF T90-008	5.5 - 8.5
Turbidity	NTU	350	NF T90-033	-
Conductivity	mS /cm	2.410	NF T90-031	-
Suspended solids	mg/l	233.89	NF EN 872	35
BOD_5	mg/l	317.03	NF EN:1899-2	30
COD	mg/l	529.08	ISO15705	125
Oxidation by $KMnO_4$	mg/l	29	NF EN ISO 8467	-

3.3. Coagulation experiments

Coagulation and flocculation processes proved to be essential in a number of diverse disciplines, including raw water and wastewater treatment. In this study, these two procedures were carried out using the jar-test apparatus. To do this, the coagulant was first injected at the beginning of the rapid mixing phase, at the speed of 150 rpm, for 10 min, 20 min, 30 min and 1 h to ensure complete dispersion, followed by slow stirring, at 50 rpm, for 10 min. It should be noted that all the experiments were conducted at neutral pH. Moreover, aluminum chloride was selected to be used as a coagulant in the chemical treatment.

Afterwards, the coagulant was introduced into wastewater with different amounts (0.1, 0.2, 0.4, 0.6, 0.8, 1, 1.2, 1.6 and 2 g). In the next step, the samples were allowed to decant for a period ranging between 30 and 60 minutes. It should be noted that flakes were observed during and after agitation. After decantation, some samples were taken from each beaker in order to measure physical parameters such as the pH, turbidity, COD and BOD_5. The coagulant dose was chosen so as to obtain the highest percentages of removal of each of the above mentioned parameters.

4. Results and discussions

4.1. Characterization of the bioflocculant potato peel

Figure 1 clearly shows the FTIR spectra of the potato peel powder. The FTIR spectrum of potato peel exhibits a broad band around 3421.5 cm^{-1} due to the presence of hydroxyl groups in cellulose, lignin and adsorbed water **[31].** Another band, recorded at 2925.8 cm^{-1}, is attributed to C-H vibrations of the CH_2 group. With regard to the band at 1647.1 cm^{-1}, it corresponds to the stretching vibrations of the C = C carbonyl functions.

In addition, close examination of the spectrum reveals the presence of two bands at 1537.2 and 1548.7 cm^{-1} which are characteristic of the deformation of the C-H bond vibration. On the other hand, several bands are observed between 1388 and 1157.2 cm^{-1}, which confirms the lignin structure of the material. Finally, another peak was noted at 1026 cm^{-1}; it is attributed to the C-O-C elongations of the cellulose functions present in the potato peel **[32].**

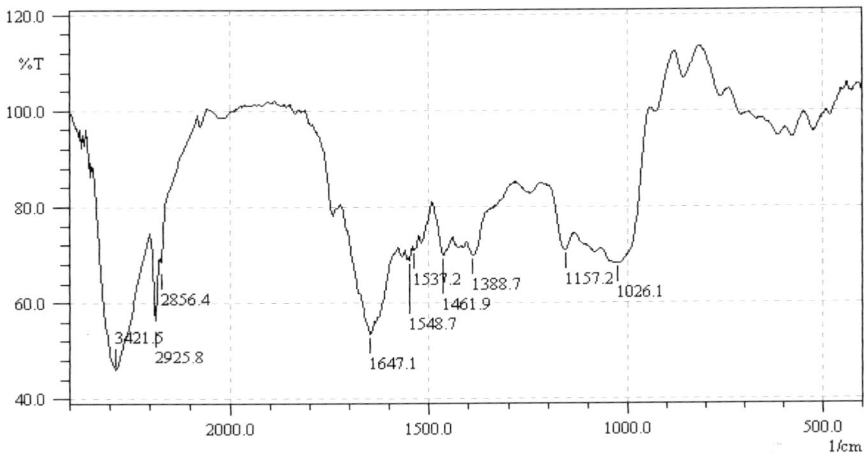

Figure 1. FTIR spectra of potato peel

4.2. Physicochemical characteristics of wastewater

The results of the physicochemical analyses of wastewater before treatment are summarized in table 1. It can be seen that the pH is equal to 8.12. It is worth also noting that the wastewater under study has an unpleasant odor. This bad odor is probably due to the presence of some chemicals or decomposing organic matter, or may be to Clostridium perfringens that give off hydrogen sulfide (H_2S). One may clearly note that the chemical oxygen demand (COD) and biochemical oxygen demand (BOD_5) values are much higher than those of other pollutants. Further, the results show that the pollution level of this wastewater is quite high; the turbidity is equal to 350 UNT as a result of the high presence of solid materials, especially suspended solids. It is important to mention that this parameter is essential in the coagulation processes primarily because it is easy to measure and is quite sensitive to coagulation processes.

4.3. Coagulation-flocculation treatment method

4.3.1. Optimization of coagulant dosage

The coagulant used in this study is aluminum chloride ($AlCl_3$), with doses ranging from 0.1 to 2g. Also, the variation of the turbidity rate (%) in terms of doses of coagulant ($AlCl_3$) is presented in

Figure 2 in order to determine the optimum coagulant dose. The same figure depicts the coagulation performance of aluminum chloride in terms of turbidity reduction of the wastewater samples.

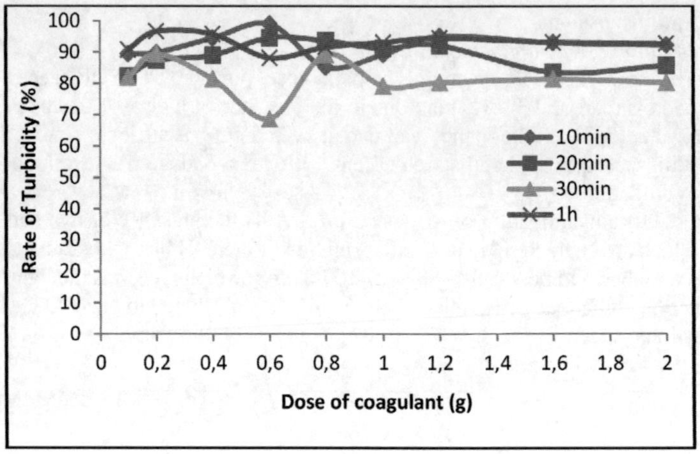

Figure 2. Evolution of the turbidity rate as a function of the dose of the coagulant aluminum chloride

This figure indicates that the turbidity continuously decreases as the quantity of the coagulant added increases. It is observed that turbidity starts decreasing after the addition of 0.1 g of aluminum chloride; it then continues declining. It should be noted that the optimal coagulant dose can be determined by measuring the turbidity and also considering the stirring time. On the basis of the results obtained, it can be seen that in the absence of flocculant, the mass of 0.6 g of aluminum chloride coagulant allows achieving a very good flocculation with a turbidity rate equal to 99.01% after for a stirring period of 10 mn. This represents the first step in wastewater treatment. The next step consists in adding the flocculant.

4.3.2. Effect of pH on coagulation efficiency

It is broadly acknowledged that the pH is a fundamental parameter. Figure 3 clearly displays the evolution of the pH as a function of the dosage of aluminum chloride coagulant, for different time periods. This figure shows that an increase in aluminum chloride dosage results in a lower pH value. In addition, depending on the characteristics of the water under study, there is an optimum pH value that allows achieving the best coagulation. It should also be noted that this pH is often within the range where the solubility of the coagulant used is minimal; this should therefore give good precipitation. Consequently, the most adequate pH value, which gives the optimal turbidity rate of 4.1, leads to the best coagulation efficiency. Note also that controlling the pH allows improving coagulation when the temperature of water is low **[33].**

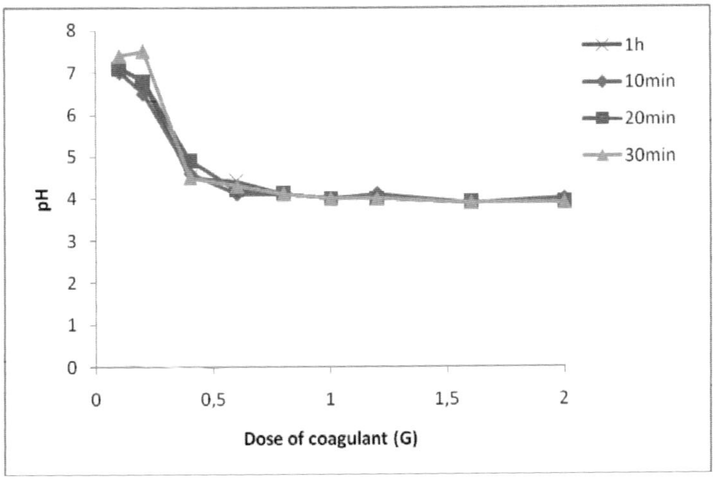

Figure 3. Evolution of pH as a function of the dose of aluminum chloride coagulant

4.3.3. Influence of the bioflocculant dosage on the coagulation-flocculation treatment

The coagulation-flocculation treatment was carried out through the addition of a constant dose of aluminum chloride (0.6g) to wastewater samples kept under continuous stirring. Next, various masses of potato peel powder, ranging between 0.1 and 1 g, were added to the resulting coagulated product. The results obtained suggested that potato peel powder with the optimum dose of 0.2g allowed flocculating wastewater efficiently, with a turbidity reduction of 98% (Figure 4). In addition, the BOD_5 dropped from 317.03 mg of O_2 / l to 0 mg of O_2 / l, which represents a 100 % reduction.

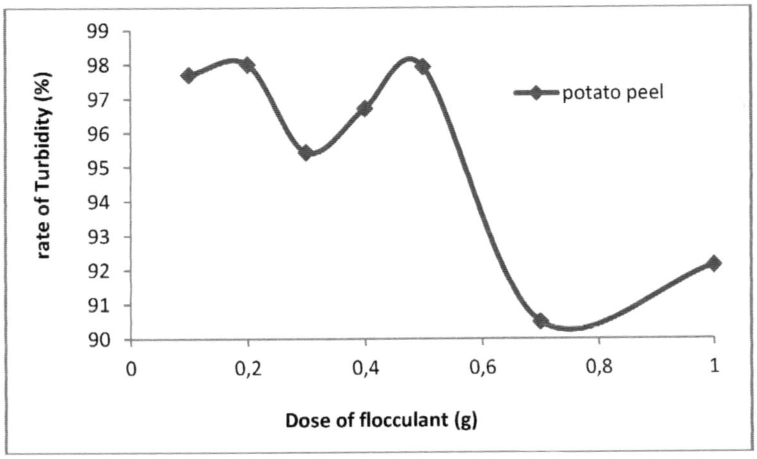

Figure 4. Evolution of turbidity rate as a function of bioflocculant dosage.

The pH is another critical factor affecting the coagulation-flocculation process. Aluminum chloride is among the most common coagulants that are used to achieve flocculation and clarification of water as well. The combination of the aluminum chloride coagulant with potato peels results in a slight drop, followed by a rise and then a decrease in pH (Figure 5). This fluctuation is primarily due to the hydrolysis of the salts used.

.

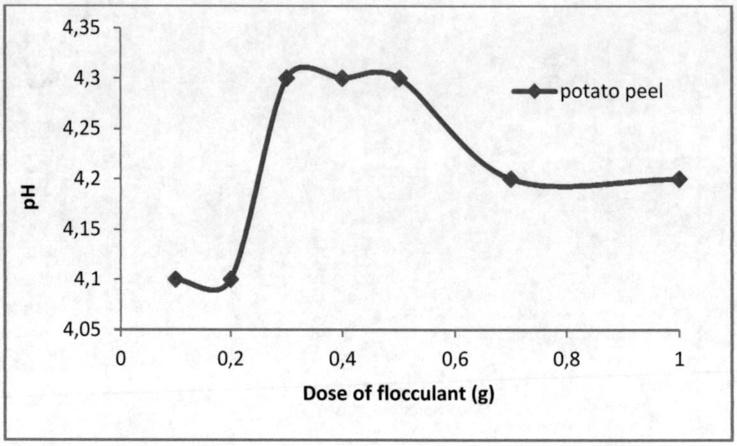

Figure 5 . Evolution of pH of wastewater as a function of bioflocculant dosage, in the presence of coagulant

5. Conclusions

This work was an attempt to recover and valorize potato peel waste, which is largely available, to be used as a low cost flocculation agent for the treatment of wastewater in the City of Saida through the coagulation-flocculation process.

The results obtained indicated that the optimum dose of aluminum chloride coagulant of 0.6 g allowed achieving high turbidity removal efficiency (99.01%) for a contact time of 10 mn. The bioflocculant dosage used was quite efficient in reducing the suspended solids, color, COD, BOD_5 and turbidity of wastewater. The findings confirmed the successful flocculation process, which led to a 100% BOD_5 reduction. It was therefore concluded that the coagulation-flocculation process can be viewed as a viable technique for the treatment of wastewater.

References

[1] Said Benkhayaa , Souad M'rabetb and Ahmed ElHarf 2020, A review on classifications, recent synthesis and applications of textile dyes, *Inorg. Chem. Commun*. **115** 107891.

[2] Rachid Hsissoua, Bouchra Benzidiab , Najat Hajjajib and Ahmed Elharfia 2018, Elaboration and electrochemical studies of the coating behavior of a new nanofunctional epoxy polymer on E24 steel in 3.5 % NaCl, *Portugaliae Electrochimica Acta*, **36** (4), 259-270.

[3] Rachid Hsissoua, Bouchra Benzidiab, Najat Hajjajib and Ahmed Elharfia 2019, Elaboration, electrochemical investigation and morphological study of the coating behavior of a new polymeric polyepoxide architecture: crosslinked and hybrid decaglycidyl of phosphorus pentamethylene dianiline on E24 carbon steel in 3.5% NaCl, Portugal. Electrochim. Acta **37**, 179-191.

[4] Rachid Hsissou, Said Abbout, Rajaa Seghiri, Malak Rehioui, Avni Berisha, Hamid Erramli , Mohammed Assouag and Ahmed Elharfi 2020, Evaluation of corrosion inhibition performance of phosphorus polymer for carbon steel in [1M] HCl: computational studies (DFT, MC and MD simulations), *J .Mater. Res. Technol,* **9** (3), 2691-2703.

[5] Hanane Arroub, Rachid Hsissou and Ahmed Elharfi 2020, Investigation of modified chitosan as potential polyelectrolyte polymer and eco-friendly for the treatment of galvanization wastewater using novel hybrid process, *Results in Chemistry* **2**, 100047.

[6] Giusy Lofrano 2012, *emerging compounds removal from wastewater, Natural and Solar Based Treatments,* Springer, Netherlands.

[7] Sabino De Gisi, Giusy Lofrano, Mariangela Grassi and Michele Notarnicola 2016. Characteristics and adsorption capacities of low-cost sorbents for wastewater treatment: *a review. Sustain. Mater. Technol*, **9**, 10-40.

[8] Bao-Yu Gao, Qin-YanYue, Yan Wang and Wei-Zhi Zhou 2007, Color removal from dye-containing wastewater by magnesium chloride, *J. Environ. Manage.* **82**, 167-172.

[9] Faisal Ibney Hai, Kazuo Yamamoto and Kensuke Fukushi 2007, Hybrid treatment systems for dye wastewaters. *Crit. Rev. Environ. Sci. Technol*, **37**, 315-377.

[10] Fatima Zohra Choumane, Belkacem Benguella, B Maachou and N Saadi 2017, Valorisation of a bioflocculant and hydroxyapatites as coagulation-flocculation adjuvants in wastewater treatment of the steppe in the wilaya of Saida (Algeria), *Ecological Engineering*, **107** ,152-159.

[11] John Bratby 2008, *Coagulation and Flocculation in Water and Wastewater Treatment*, second ed, IWA publishing, London.

[12] Pang FongMoi, Teng SheauPing, Teng TjoonTow and Omar A K M 2009, Heavy metal removal by hydroxide precipitation and coagulation-flocculation methods from aqueous solutions, *Water Qual. Res. J.* **Can. 44** (2) 174-182.

[13] Chai Siah Lee, John Robinson and Mei Fong Chong 2014, A review on application of flocculants in wastewater treatment, *Process Saf. Environ. Prot.* **92**, 489-508.

[14] Victor Ajao, Remco Fokkink, Frans Leermakers, Harry Bruning, Huub Rijnaarts and Hardy Temmink 2021, Bioflocculants from wastewater: Insights into adsorption affinity, flocculation mechanisms and mixed particle flocculation based on biopolymer size-fractionation, *Journal of Colloid and Interface Science,* **581**, 533-544.

[15] T Nharingo and M Moyo 2016. Application of Opuntia Ficus-Indica in Bioremediation of Wastewaters. A Critical. Review. *J. Environ. Manage*, **166**, 55-72.

[16] Lorenzo A Picos-Corrales, Juan I Sarmiento-Sanchez, Jose P Ruelas-Leyva, Gregorio Crini, Eduardo Hermosillo-Ochoa and J Ariel Gutierrez-Montes 2020, Environment-friendly approach toward the treatment of raw agricultural wastewater and river water via flocculation using chitosan and bean straw flour as bioflocculants, *American Chemical Society (ACS) Omega*, **5**, 3943-3951.

[17] Grégorio Crini, Eric Lichtfouse, Lee Wilson and Nadia Morin-Crini 2019. Conventional and Non-Conventional Adsorbents for Wastewater Treatment. *Environ. Chem. Lett*, **17**, 195-213.

[18] A Mishra and M Bajpai 2005, Flocculation behaviour of model textile wastewater treated with a food grade polysaccharide, *J. Hazard. Mater.* **14** (118) 213-217.

[19] T Suopajärvi, H Liimatainen, O Hormi and J Niinimäki 2013, Coagulation-flocculation treatment of municipal wastewater based on anionized nanocelluloses, *Chem. Eng. J.* **231**,59-67.

[20] C Wu, Y Wang, W Zhao, Z Yanxia and Q Yue 2012, Coagulation performance and floc characteristics of aluminum sulfate using sodium alginate as coagulant aid for synthetic dying wastewater treatment, *Sep. Purif. Technol.* **95**,180-187.

[21] V Ajao, H Bruning, H Rijnaarts and H Temmink 2018, Natural flocculants from fresh and saline wastewater: Comparative properties and flocculation performances, *Chem. Eng. J.* **349**, 622-632.

[22] F Renault, B Sancey, J Charles, N Morin-Crini, PM Badot, P Winterton and G. Crini 2009, Chitosan flocculation of cardboard-mill secondary biological wastewater, *Chem. Eng. J.* **155**,775-783.

[23] H Y Gebrechristos and W Chen 2018, Utilization of potato peel as eco-friendly products: A review. *J. Food Sci and Nutr*, **6** (6), 1352-1356.

[24] S Boumchita, A. Lahrichi, Y Benjelloun, S Lairini, V Nenov and F Zerrouq 2016, Removal of cationic dye from aqueous solution by a food waste: potato peel. *J. Mater. Environ. Sci.* **7**(1), 73-84.

[25] Mohini Verma and R Naresh Kumar 2016, Can coagulation-flocculation be an effective pre-treatment option for landfill leachate and municipal wastewater co-treatment?, *Perspectives in Science*, **8**, 492-494.

[26] AFS Maldonado, E Mudge, MG Gänzle and A Schieber 2014, Extraction and fractionation of phenolic acids and glycoalkaloids from potato peels using acidified water/ethanol-based solvents. *Food Res Int*, **65**, 27-34.

[27] KB Jeddou, F Chaari, S Maktouf, O Nouri-Ellouz , CB Helbert and R E Ghorbel 2016, Structural, functional, and antioxidant properties of water-soluble polysaccharides from potatoes peels. *Food Chem*, **205**, 97-105.

[28] S Liang and AG McDonald 2014, Chemical and thermal characterization of potato peel waste and its fermentation residue as potential resources for biofuel and bioproducts production. *J Agric Food Chem,* **62**, 8421-8429.

[29] S Liang, AG McDonald and ER Coats 2014, Lactic acid production with undefined mixed culture fermentation of potato peel waste. *Waste Manag*, **34**, 2022-2027.

[30] Ahsan Javed, Awais Ahmad, Ali Tahir, Umair Shabbir, Muhammad Nouman and Adeela Hameed 2019, Potato peel waste-its nutraceutical, industrial and biotechnological applacations", *AIMS Agriculture and Food*, **4**(3),807-823.

[31] N Wibowo, L Setyadhi, D Wibowo, J Setiawan and S Ismadji 2007. Adsorption of benzene and toluene from aqueous solution onto activated carbon and its acid heat treated forms: Influence of surface chemistry on adsorption. *Journal of Hazardous Materials*, **146**, 237-242.

[32] K Bouhadjra, W Lemlikchi and N Oubagha 2017, Application of potato peels for the removal of reactive blue 72 from aqueous solutions, J. *Wat. Env. Sci*, Vol **1**, (Numéro spécial ICWR 2), 219-229.

[33] R JULIEN1983, *Etude de la coagulation-floculation-décantation d'une eau colorée froide de faible alcalinité.* Mémoire de maîtrise - Département de Génie Civil, École Polytechnique de Montréal, Canada, 157p.

IOP Conf. Series: Materials Science and Engineering 1204 (2021) 012003 doi:10.1088/1757-899X/1204/1/012003

Nonlinear regression for analysis of the uptake of KMnO4 from aqueous solutions onto natural marl

E A Bouzeghaia*, S Hafsi, N Gherraf

Department of matter sciences, Larbi Ben M'hidi University, Oum El Bouaghi, 04000 Algeria

atra.bouzeghaia@yahoo.com

Abstract. The present work aims to evaluate the performance of raw marl collected from the region of Oum El Bouaghi on the elimination of $KMnO_4$ from aqueous solution. The adsorption capacities were studied using the batch technique as a function of pH, initial concentration and temperature. The maximum adsorption for $KMnO_4$ was found to be 83.517% at pH 3.0. When the pH is low (pH<3.0), the uptake of $KMnO_4$ occurs with an electrostatic attraction between the MnO_4^- ions and the surface proton groups, while at pH $\geq$3.0-8.0, the formation of complexes with the surface functional groups during the adsorption process and/or cation exchange are responsible for retention of Mn(II) ions onto marl. Experimental adsorption data were modeled with different nonlinear isotherms and kinetic models. Furthermore, statistical errors (SSE, SD, ARE, χ^2, R^2 and R^2_{adj}) were calculated to choose the suitable model. Langmuir isotherm was found to be best for fitting the experimental data. The calculated thermodynamic parameters indicated that adsorption process was physical, spontaneous and endothermic in nature. From this work, it was concluded that the natural marl can be used for effective elimination of $KMnO_4$ from aqueous solution and wastewater.

1. Introduction

Manganese is the second most abundant heavy metal in nature. It is used in the manufacture of iron and steel alloys, glass, welding works, fireworks, dry battery cells, electric coils and pesticides [1-3]. Manganese can exist in aqueous solutions with different oxidation states (II to VII). The water soluble forms are Mn(II) and Mn(VII) [4]. Manganese is considered as an essential micronutrient for organisms and plants. When manganese concentrations exceed 0.1 mg.L^{-1} in water bodies, dissolved oxygen is reduced, giving off an undesirable taste that can affect the central nervous system of living organisms [5, 1]. Due to the harmful effects of manganese, its ions removal from wastewater and industrial effluents is crucial. Among the physico-chemical water treatment processes, the adsorption method is commonly used because it is efficient, simple, selective and inexpensive. Recently, the search for new efficient, inexpensive and eco-friendly adsorbents is gaining an increasing interest by researchers. Regarding Marl, it is a sedimentary rock composed of a mixture of clay and carbonate [6]. However, the data available on the adsorption of toxic metals on this type of material is limited. The aim of this study, thus, is to assess the feasibility of using raw marl from the Oum El Bouaghi region as a new natural adsorbent to remove MnO_4^- from the aqueous solution. The study of the reaction parameters (initial concentration, pH and temperature) and the adsorption isotherms were carried out in a batch

IOP Conf. Series: Materials Science and Engineering 1204 (2021) 012003 doi:10.1088/1757-899X/1204/1/012003

system. Modeling of adsorption isotherms was estimated using nonlinear regression. The best models are selected after analyzing the errors. The thermodynamics of adsorption were also carried out in this work.

2. Experimental

The adsorbent used is marl taken from the R'ghiss Mountain located in the region of Oum El Bouaghi (Algeria). It is collected, washed with distilled water, dried in the open air and sieved through a 180 μm sieve. Mineralogical analysis has shown that this solid is composed of 15% quartz, 17% calcite, 05% albite, 23% illite, 13% kaolinite, 20% montmorillonite and 07% ferruginous minerals + others. Its specific surface determined by the BET method is 14.603 m².g^{-1} [7]. The stock solution of MnO_4^- with a concentration of 200 mg.L^{-1} was prepared by dissolving potassium permanganate ($KMnO_4$, M = 158.04 g.mol^{-1}, purity: 99%, Merck) in distilled water. The solutions at different concentrations used in our study were obtained by diluting the stock solution. The pH of the solution is adjusted by adding 0.1 M HCl or 0.1 M NaOH solutions. The pH value of the solution was measured by a Jenway model pH meter. The adsorption experiments were carried out by the batch method by bringing 25mL of different experimental solutions of MnO_4^- into contact with 0.4g of marl in 500ml conical flasks. The effects of initial adsorbate concentration (4–10 mg.L^{-1}), pH (2.0–8.0) and temperature (293–303K) on manganese ion retention by marl were investigated. The samples were stirred at a speed of 475 rpm in a water bath shaker (Heizung model) for a known period of time. Then the mixtures were centrifuged at 3500 rpm for 5 min in a Hettich Universal centrifuge. The supernatants were collected and analyzed by molecular absorption spectrophotometry (Spectronic 70 spectrophotometer). The residual concentration of manganese in aqueous solution was determined by referring to the calibration curve obtained from the test solutions at the wavelength λmax = 525 nm (figure 1). The percentage of removal (R%) of manganese ions was calculated by equation (1):

$$R(\%) = (C_0 - C_t)\frac{100}{C_0} \qquad (1)$$

where:
C_0 (mg.L^{-1}) and C_t (mg.L^{-1}) are the initial and residual concentrations of MnO_4^-.

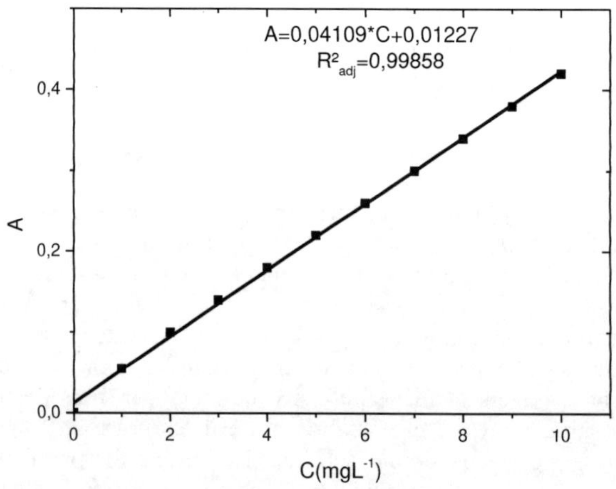

Figure 1. Calibration curve for the $KMnO_4$ solution.

2.1. Statistical analysis

In our work, the fitting of the experimental data to the isothermal models was performed according to the nonlinear regression using the Origin 9.0 software. Nonlinear regression is the best method to determine the parameters of isotherms [8]. Six different error functions (SSE, SD, χ^2, ARE, R^2, and R^2_{adj}) were analyzed to examine the validity of mathematical models and predict the best models.
The error functions are presented by the following equations:

- The Sum of Squared Errors (SSE)

$$SSE = \sum_{i=1}^{N} (q_{t,cal} - q_{t,exp})_i^2 \quad (2)$$

- Standard deviation (SD)

$$SD = \left(\frac{1}{N-P} \sum_{i=1}^{N} \left(q_{t,cal} - q_{t,exp} \right)_i^2 \right)^{1/2} \quad (3)$$

- Chi-square test (χ^2)

$$\chi^2 = \sum_{i=1}^{N} \left[\frac{\left(q_{t,exp} - q_{t,cal} \right)^2}{q_{t,cal}} \right]_i \quad (4)$$

- Average relative error (ARE)

$$ARE = \frac{100}{N} \sum_{i=1}^{N} \left| \frac{q_{t,cal} - q_{t,exp}}{q_{t,exp}} \right|_i \quad (5)$$

- Coefficient of determination (R^2)

$$R^2 = \frac{\left(q_{t,exp} - \overline{q_{t,cal}} \right)_i^2}{\sum_{i=1}^{N} \left(q_{t,exp} - \overline{q_{t,cal}} \right)^2 + \left(q_{t,exp} - q_{t,cal} \right)^2} \quad (6)$$

- Adjusted R-squared (R^2 adj)

$$R^2_{adj} = 1 - \frac{(N-1)(1-R^2)}{(N-P)} \quad (7)$$

where:
$q_{t,cal}$ ($mg.g^{-1}$) and $q_{t,exp}$ ($mg.g^{-1}$) are the calculated and experimental adsorption amounts respectively, $\overline{q_{t,cal}}$ ($mg.g^{-1}$) refers to the average of the calculated adsorbed amount, N is the number of experimental data points, P is the number of parameters of the fitted model.

The experimental results agree with the theoretical models when the values of SEE, χ^2 SD, ARE tend towards 0, and the coefficients R^2 and R^2adj tend towards 1.

3. Results and Discussion

3.1. Effect of the initial concentration

The effect of the initial concentration on the adsorption efficiency was studied. The experimental data are represented graphically in figure 2. The retention of manganese increases strongly in the initial stages and then gradually to reach equilibrium after 180 min of the contact time. The increase in the initial

IOP Conf. Series: Materials Science and Engineering 1204 (2021) 012003 doi:10.1088/1757-899X/1204/1/012003

$KMnO_4$ concentration from 4 mg.L^{-1} to 10 mg.L^{-1} leads to an increase in the amount adsorbed from 40.539% to 83.517%. This behavior can be explained by the increase in the driving force which overcomes the various mass transfer resistances of metal ions from the aqueous phase to the solid phase. Consequently, the number of collisions between the MnO_4^- ions and the active sites of the adsorbent increases, this leads to the increase of the adsorbed quantity.

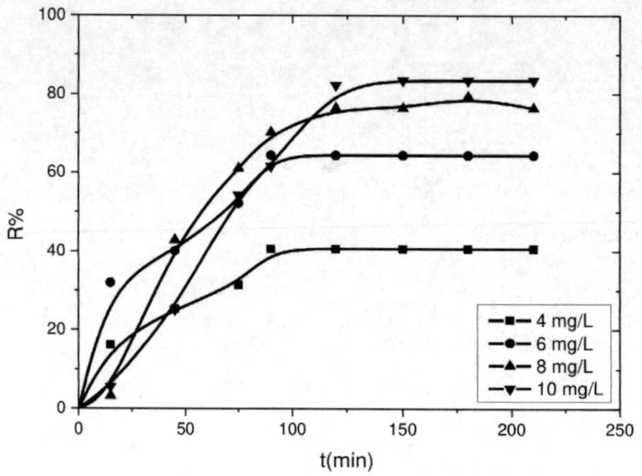

Figure 2. Effect of the initial concentration of MnO_4^-
(pH=3.0, T=293K, v=475 rpm, m=0.4g).

3.2. *Effect of pH*

Adsorption is highly dependent on pH of the aqueous solution since it is attributed to distribution of different adsorbed species in the medium as well as the surface charge and degree of adsorbent ionization [9]. Manganese adsorption was carried out in acidic, neutral and basic medium in the pH range 2 to 8 as shown in figure 3. In medium of high acidity, $KMnO_4$ produces manganese (II). On the other hand, $KMnO_4$ produces manganese dioxides under acid-neutral and alkaline pH conditions, as illustrated by equations (8), (9), and (10) [10].

In a strong acidity medium:

$$MnO_4^- + 8H^+ + 5é \rightarrow Mn^{2+} + 4H_2O, \ (E° = +1.51V) \quad (8)$$

In an acid-neutral medium:

$$MnO_4^- + 4H^+ + 3é \rightarrow MnO_2(s) + 2H_2O \ , (E° = +1.70V) \ (9)$$

In alkaline medium:

$$MnO_4^- + 2H_2O + 3é \rightarrow MnO_2(s) + 4OH^- \ , (E° = +0.59V) \ (10)$$

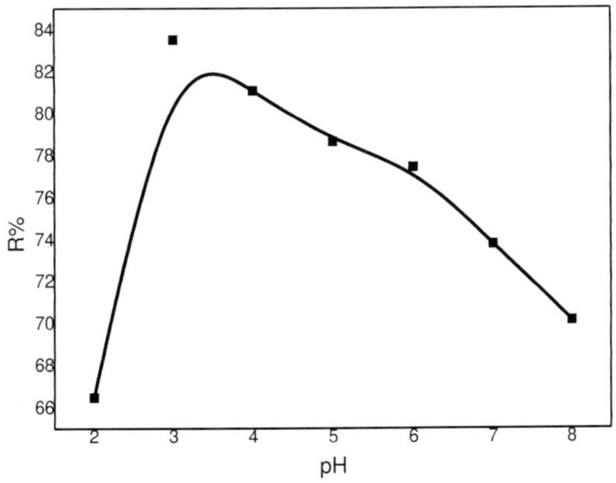

Figure 3. Effect of pH on the adsorption of $KMnO_4$ by marl (C_o=10 mg.L^{-1}, T=293K, v=475rpm, m=0.4g).

The results clearly show that the percentage of manganese extraction in aqueous solution increases up to pH 3.0 (83.517%). In this case, the elimination of $KMnO_4$ by the marl refers to the electrostatic attraction between the MnO_4^- ions and the protonated groups of the surface involving the formation of an ion pairing and favoring the elimination of Mn^{7+} in aqueous solution. In addition, we can also observe that the removal rate of manganese decrease from 83.517% to 70.131% as pH rises above 3.0 due to the increase of the electrostatic repulsion forces between the MnO_4^- and the active sites on the surface which reduces the elimination of Mn^{7+}.The deprotonating of the surface decreases the competition between the H^+ ions and the Mn^{2+} ions leading to a decrease in electrostatic repulsions. Manganese (II) ions resulting from the reduction of MnO_4^- are adsorbed by the marl through the formation of complexes with active surface functional groups (OH) of complexes [11] and/or by cation exchange with cations located in exchangeable sites of clay [12]. The maximum adsorption of $Mn^{2+/7+}$ ions was found at pH 3.0 and therefore the optimum pH value is 3.0.

3.3. *Effect of Temperature*

The study of the evolution of MnO_4^- elimination was carried out at different temperatures: 293K, 298K, 303K (figure 4). We perceive an increase in $Mn^{2+/7+}$ ion extraction from 83.51667% to 88.38404% with increasing temperature, revealing the endothermic nature of adsorption. In contrast, the increase in adsorption capacity can be attributed to the acquisition of additional kinetic energy to facilitate the diffusion and removal of Mn.

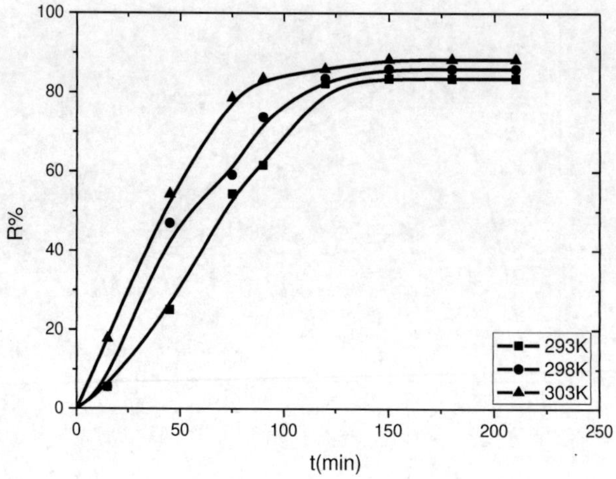

Figure 4. Effect of temperature effect (pH=3.0, C_o=10 mg.L^{-1}, v=475rpm, m=0.4g).

3.4. *Adsorption Isotherms*

In our study, the experimental data corresponding to the equilibrium are analyzed in the temperature range 293K to 303K at pH=3.0 by isotherms models Langmuir (equation (11)) [13] and Freundlich (equation (12)) [14].

$$q_e = \frac{q_m K_L C_e}{1 + K_L C_e} \quad (11)$$

$$q_e = K_F C_e^{1/n} \quad (12)$$

where:

K_L (L.mg^{-1}) is the equilibrium constant of Langmuir isotherm, q_m (mg.g^{-1}) is the maximum adsorption amount of the solute on the solid phase, q_e is the amount of adsorbed of the solute on the solid phase at equilibrium (mg.g^{-1}), C_e(mg.L^{-1}) is the concentration of the metal in the solution at equilibrium, K_F is the constant of the Freundlich isotherm, n expresses the adsorption intensity (without units).

Table 1 represents the various parameters calculated from the graphic representation of these two models. A slight improvement in manganese adsorption was found with increasing temperature, which shows that the adsorption process is endothermic. Similar results are found by Kebabi et al (2017) [15]. The Freundlich parameter values (n) ($1.894 \leq n \leq 2.459$) indicate that the adsorption of Mn on the marl is favorable, the exponent n gives an indication of the adsorption favorability. The values of K_L and q_m increase with the rise of temperature, which means that the affinity between manganese ions and marl develops with increasing temperature and the adsorption of $Mn^{2+/7+}$ions on the marl has an endothermic nature. Note that, the Langmuir isotherm model gives the best fit of the experimental data with values of R^2 and R^2_{adj} closer to unity and of order greater than 0.98. The values of SSE, SD, χ^2, ARE corresponding to the Langmuir isotherm are lower than those obtained by the Freundlich isotherm. This result is consistent with other works in the literature [16-19].

Table 1. Parameters of ion adsorption isotherms of $Mn^{2+/7+}$ on marl.

		T(K)		
Isotherm	Parameter	293	298	303
	q_m(mg.g^{-1})	0.512	0.55	0,585
	K_L(L.mg^{-1})	0.903	0.974	1,103
	SSE	$8.92.10^{-4}$	$5.54.10^{-4}$	$2.716.10^{-4}$
	SD	0.014	0.012	0.008
Langmuir	χ^2	$31.5.10^{-4}$	$18.2.10^{-4}$	$7.904.10^{-4}$
	ARE	4.58	3.33	1.835
	R^2	0.99	0.994	0.997
	R^2_{adj}	0.987	0.993	0.997
	K_F (mg$^{(1-n)}$.L^n.g^{-1})	0.243	0.268	0.3
	1/n	0.407	0.421	0.43
	n	2.459	2.374	1.894
	SSE	$15.7.10^{-4}$	$12.8.10^{-4}$	$10.3.10^{-4}$
Freundlich	SD	0.023	0.021	0.018
	χ^2	$55.5.10^{-4}$	$42.7.10^{-4}$	$32.2.10^{-4}$
	ARE	5.618	4.845	4.156
	R^2	0.912	0.939	0.958
	R^2_{adj}	0.883	0.919	0.944

3.5. *Thermodynamic study*

Thermodynamic parameters such as Gibbs free energy change (ΔG°), enthalpy change (ΔH°) and entropy change (ΔS°) are calculated from the following equations [20]:

$$\Delta G^0 = -RT\ln(K_e^0) \quad (13)$$

$$\Delta G^0 = \Delta H^0 - T\Delta S^0 \quad (14)$$

$$\ln(K_e^0) = \frac{\Delta S^0}{R} - \frac{\Delta H^0}{R} x \frac{1}{T} \quad (15)$$

where:

$\mathbf{K_e^\circ}$ is the thermodynamic equilibrium constant (dimensionless).

The slope and the intercept of a graph of $\mathbf{Ln(K_e^\circ)}$ vs (1/T) are used to determine the values of ΔH° and ΔS° respectively (figure 5). The results are collated in table 2.

The results reveal the decrease in ΔG° from -28.244 to -29.714 kJ.mol^{-1}, these negative values indicate the feasibility and spontaneity of the adsorption process of $Mn^{2+/7+}$ ions. The positive value of ΔH° (14.806 kJ.mol^{-1}) shows that the adsorption process is physical and endothermic in nature. The positive value of ΔS° (146.857 J.K^{-1}.mol^{-1}) provides evidence that the adsorption process has increasing random behavior at the solid - solution interface.

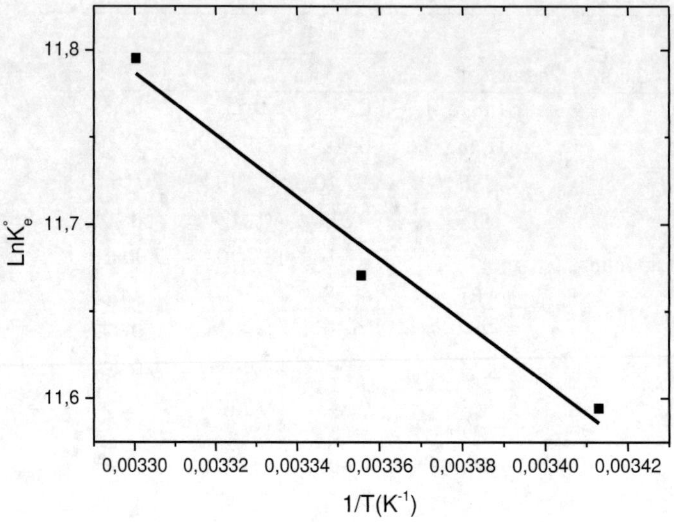

Figure 5. Representation of the variation of Ln(K_e°) as a function of 1/T.

Table 2. Thermodynamic parameters of adsorption of $Mn^{2+/7+}$ by marl.

T(K)	ΔG° (kJ.mol^{-1})	ΔS° (J.K^{-1}.mol^{-1})	ΔH° (kJ.mol^{-1})	R^2_{adj}
293	-28.244			
298	-28.915	146.857	14.806	0.957
303	-29.714			

4. Conclusion

In this study, marl of Oum El Bouaghi region was used as an adsorbent to remove MnO_4^- from the aqueous solution. The main results are as follows:

- The initial concentration and temperature have positively influence the removal of $Mn^{2+/7+}$ ions.
- The adsorption process reaches equilibrium after 180 minutes.
- Maximum removal of Mn was found at pH = 3.0. When the pH is below 3.0, the percentage of extraction of manganese in aqueous solution increases. The electrostatic attraction between the MnO_4^- ions and the proton groups on the surface. When the pH is above 3.0, the increase in electrostatic repulsive forces between MnO_4^- ions and active sites on the surface decreases the removal of Mn^{7+} ions. Mn^{2+} ions can be adsorbed on the marl by the formation of complexes with the active surface functional groups (OH) of the complexes and/or by cation exchange with the cations located in the exchangeable sites of the clay.
- The equilibrium data were well fitted by the Langmuir isotherm.
- The thermodynamic study shows that the adsorption process is physical, endothermic, and spontaneous.

The results of this study support the use of Oum El Bouaghi marl to effectively remove $KMnO_4$ from wastewater.

References

[1] Mahadevaiah N, Vijayakumar B, Hemalatha K and Prakash B J 2011 *Bull. Mater. Sci.* **34** 1675
[2] Winder C 2004 Toxicity of metals *Occupational Toxicology* vol 2,ed C Winder and N H Stacey (Boca Raton: CRC Press) chapter 12 p 298
[3] Nieva A D, Doma B T and Chao H P 2018 *IOP Conf. Ser.: Earth Environ. Sci.* **191** 012034
[4] Patil D S, Chavan S M and Oubagaranadin J U K 2016 *J. Environ. Chem. Eng.* **4** 468
[5] Chen F, Hong M, You W, Li C and Yu Y 2015 *Appl. Surf. Sci.* **357** 856
[6] Cornell R M and Aksoyoglu E S 1991 *Clay. Miner.* **26** 567
[7] Bouzeghaia E A, Hafsi S, Gherraf N 2018 *World. J. Environ. Biosci.***7** 8
[8] Jasper E E, Ajibola V O and Onwuka J C 2020 *Appl. Water. Sci.* **10** 1
[9] Anirudhan T S, Nima J and Divya P L 2013 *Appl. Surf. Sci.* **279** 441
[10] Liu R, Liu H, Zhao X, Qu J and Zhang R 2010 *J. Hazard. Mater.* **176** 926
[11] Mahmoud M E, Yakout A A, Saad S R, and Osman M M 2016 *Desalination. Water. Treat.* **57** 15559
[12] Eba F, Gueu S, Eya'A-Mvongbote A, Ondo J A, Yao B K, Ndong N J and Kouya B R 2010 *Int. J. Eng. Sci. Technol.* **2** 5001
[13] Langmuir I 1918 *J. Am. Chem. Soc.* **40** 1361
[14] Freundlich H 1906 *J. Phys. Chem.* **57** 385
[15] Kebabi B, Terchi S, Bougherara H, Reinert L and Duclaux L 2017 *Appl. Clay. Sci.* **139** 92
[16] Taffarel S R and Rubio J 2009 *Miner. Eng.* **22** 336
[17] Kocaoba S 2009 *Desalination.* **244** 24
[18] AL-Aoh H A 2019 *Mater. Res. Express.* 6 115102
[19] Alcaraz L, Alguacil F J and López F A 2021 *Can. J. Chem. Eng.* **99** 447
[20] Lima E C, Hosseini-Bandegharaei A, Moreno-Piraján J C and Anastopoulos I 2019 *J. Mol. Liq.* **273** 425

Chemiresistive sensors based on Dodecyl benzene Sulfonic acid doped Polypyrrole and Reduced Graphene Oxide for nitrogen oxides

Djamil GUETTICHE[1], Ahmed MEKKI[1], TIGHILT Fatma Zohra[2], Noureddine RAMDANI[1], Rachid MAHMOUD[1]

[1]Ecole Militaire Polytechnique, BP 17, Bordj El Bahri 16111, Alger, Algérie
[2]Centre de Recherche en Technologie des Semi-conducteurs pour l'Energétique, Bd. Dr Frantz FANON, B.P 140 Alger-7 Merveilles 16038, Algérie.

guettiche.djamil@gmail.com

Abstract. A new series of polypyrrole doped with n-dodecylbenzene sulphonic acid /reduced graphene oxide (PPy-DBSA/rGO) nanocomposite was electrodeposited on Indium tin oxide coated Polyethylene terephthalate (ITO/PET) flexible substrate by electrochemical route using the chronoamperometric technique. As-prepared for testing of chemiresistive properties against the detection of nitrogen dioxide (NO_2) vapors at room temperature. The sensitivity and reactivity of the composite toward NO_2 was evaluated. The recorded morphological and structural data confirmed that the PPy-DBSA/rGO forms a homogeneous nanocomposite. The optimal NO_2 sensing properties have been revealed by the PPy-DBSA/rGO in terms of response (43%), response time (30.25 s), the detection limit (1ppm), and reproducibility. Furthermore, Results showed that the doped by sulfonic acid improved both the sensitivity and the reactivity of our produced nanocomposite toward NO_2. Due to the strong interactions between the NO_2 gas molecules and the rGO was dramatically enhanced the electronic properties of these nanocomposites. These striking characteristics of the newly developed nanocomposites make them very suitable to be used as NO_2 gas sensor.
Keywords. reduced graphene oxide, graphene nanoplatelets,Polypyrrole, NO2 sensing, nanocomposite.

1. Introduction

Nitrogen dioxide gas (NO_2) is one of the hazardous environmental pollutants that is caused by the breakdown of ester energetic elements, causing dangerous autoignition under certain conditions [1]. In addition, the rapid sensing of NO_2 vapors at a ppm level is strongly recommended to alleviate all of the aforementioned issues and expected disasters [1]. This gas is known for its high toxicity for human health and environment especially when it is contacting oxygen or water, where it can be to provoke respiratory infections that leads to susceptible death [2]. The necessity for detection of this gas at low concentrations grew due to rigorous environmental regulations and the requirement to control emissions from various activities [1].

Metal oxides based NO_2 sensors family are widely investigated, where they showed a good stability, altering oxygen stoichiometry, and develop a polarity on its surface in the presence of target gases [3]. However, due to their high operating temperature, the intrinsic resistance drift of MO-based sensors has limited their practical use. therefore, conductive polymers doped with organic surfactants have recently been developed to replace these conventional inorganic semiconductors due to their room temperature sensing ability, high selectivity, improved stability and short response time characteristics [4]. For example, filled and pure conductive polymers of polyaniline, polyphenylene, polythiophene and PEDOT have been widely used for the chemical detection of various hazardous gases.

Polypyrrole (PPy) is getting prominence because to its easy production, excellent electrical conductivity, good reduction-oxidation characteristic, and long ambient life [5]. The morphology, surface area, and electrical properties of PPy can be controlled using different dopants [6,7] such as camphor sulphonic acid (CSA) and dodecylbenzene sulfonic acid (DBSA) to improve the solubility of polymers with high conductivity. to develop novel gas sensors with lower detection limits, faster response times, and higher sensitivity.

However, carbon-based nano-fillers, particularly graphene, is widely used to reinforce PPy conductive polymer doped with sulfonic acid owing to its outstanding feature [7], Suitable for increasing gas sensing properties. One step synthesis of PPy-graphene nanocomposite doped with DBSA: (i) the exfoliation of reduce graphene oxide with DBSA, when formed electrolyte solution (ii) electrochemically deposition of PPy and graphene loaded with dopant during the electro polymerization reaction. On the other hand, several procedures were adopted for producing PPy-Graphene films nanocomposite, including the in-situ chemical approaches, drop- coating, spin-coating, and dip coating [8]. However, graphene segregation in the polymer matrix is a significant complaint. Thus, the dispersion state of graphene in organic solutions can be promoted by exfoliation using various surfactants such as n-dodecylbenzene sulphonic acid [9]. This exfoliation of reduce graphene oxide in PPy matrix can also have enhanced sensing characteristics of PPy sensors by increasing the surface area and the performance sensorial proprieties. To the aimed to contribute, no previous research has been conducted to investigate the potential of polypyrrole-reduced graphene oxide doped with DBSA nanocomposites as a NO_2 chemiresistive sensor.

Herein, we present a one-step synthesis of PPy-DBSA/reduced graphene oxide nanocomposite materials. Then, a comparison with electrochemically-polymerized PPy-DBSA/rGO nanocomposites films of NO_2 sensing has been made. In addition, the composites were also characterized by RAMAN, SEM, chronoamperometry, and conductivity. Finally, several analysis approaches were used to determine the sensor's sensitivity, reaction time, reproducibility and the effect of humidity.

2. Experimental section

2.1 Materials

Pyrrole (Sigma-Aldrich, purity>98%) was distilled under reduced pressure, and then it was stocked at 4 °C in the dark before synthesis. Organic solvents used as received, and deionized (DI) water was also used. rGO was synthesized by ourselves in the laboratory, dodecylbenzene sulfonic acid (DBSA) was purchased from Sigma Aldrich. All electrochemical measurements were performed using an Autolab 101 type device.

2.2 Methods

Electron microscopy micrographs were taken on Scanning Electron Microscope (FEI-Quanta 600). Raman spectroscopy was carried out using (Horiba Labram HR evolution), excitation was at 633 nm from He-Ne Laser at room temperature. The four-point probe technique measured the electrical conductivity of films (with a typical probe separation of 1 mm) using the Jandel RM3000 apparatus.

2.3 Electrochemical polymerization

Typically, the PPy-DBSA / rGO processed films were electrochemically deposited on indium tin oxide ITO substrates at room temperature. Using potentiostat- galvanostat type VMP3 -AUTOLAB. They were using constant potential amperometry in a three-electrode electrochemical cell. The electrochemical deposition was done at a potential of 2.5 V with 0.01 mol L^{-1} DBSA, 0.01 mol. L^{-1} pyrrole, and 0.5 g. L^{-1} of the reduce graphene oxide. In a configuration with three electrodes where the ITO substrate constitutes the working electrode, a reference electrode is Ag/AgCl, and a counter electrode is a platinum (Pt). In order to prepare the electrolyte used in the electrochemical cell in order to deposit the PPy-DBSA/rGO film, a dispersion of reduce graphene oxide in acetonitrile was ultrasonicated for 15 min, then appropriate amount of pyrrole was added and stirred well to form a colloidal solution, followed by sonication for 15 min. After that, one volume of dodecylbenzene sulfonic acid DBSA was dissolved in this solution with magnetic stirring. The addition of reduce graphene oxide were also made. Afer electrodeposition of the PPy-DBSA/rGO layer on the ITO/PET, the films were rinsed with DI water and dried under azote and ambient temperature. Cyclic voltammetry (CV) was performed in 0.1 M ($C_{16}H_{36}BF_4N$) and 0.01 M acetonitrile solutions at various potential windows and scan rates 100 mV/s.

2.4 Chemiresistive sensor system set up

All sensing tests were conducted out on a handmade platform that we had previously described in ours works [10]. The nitrogen dioxide vapors were extracted from fuming nitric acid and injected using a Hamilton Microsyringe through a small hole on the test cell.

Flexible polymer nanocomposite mounted on a movable piston and inserted into a 60 mL polyethylene tube. A four-point probe attached to a device of type Jandel RM3000 is mounted on the top of the tube for dynamic electrical conductivity measurement. Samples are inserted using a microliter syringe through a tiny hole (contact point). The environmental humidity was made purely by evaporating distilled water that was attached directly to the test compartment. Electrical conductivity measurements were taken at temperatures between 25 and 30 °C. The precise concentrations of vapors released were determined using our earlier work [11]. The response % was calculated using the following equation:

$$S = |R_g - R_a| / R_a * 100\% \quad (1)$$

When R_g is the response value of sensor thin layers under measuring gas, and R_a is the response value of fresh air.

2.5 Synthesis of PPY-DBSA/rGO thin films

Figures 2 show the evolution of the chronoamperogram recorded on ITO / BOPET electrodes. The electropolymerization time is 30 s, and the imposed potential is 2.5 V.

Figure 1. Photography of PPy-DBSA/rGO films electrodeposited on ITO/BOPET.

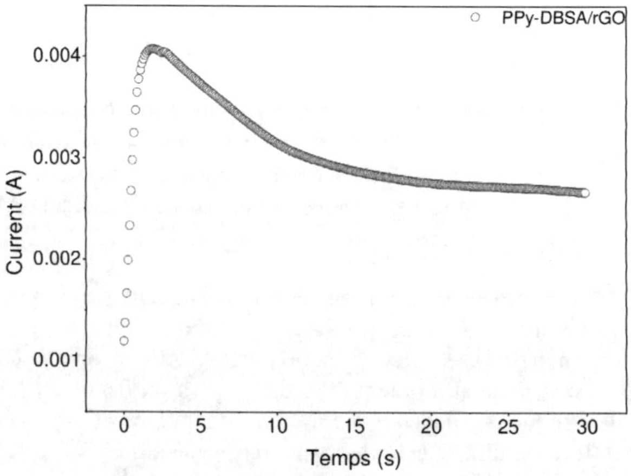

Figure 2. Chronoamperogram of an aqueous solution of 10^{-1} M pyrrole, 10^{-1} M DBSA, and 1g of rGO, on an ITO electrode (Ag / AgCl reference electrode, Pt against electrode)

IOP Conf. Series: Materials Science and Engineering 1204 (2021) 012004 doi:10.1088/1757-899X/1204/1/012004

3. Results and Discussion

3.1 Spectroscopic characterization
Typical Raman spectra of composites are shown in Fig. 3. They show bands demonstrating the characteristic peaks observed on the PPy-DBSA at 1350 cm^{-1} and 1572 cm^{-1}, which correspond to the elongation of the C = C bond of the polymer chain. Regarding pure PPy, the bands were observed in 1388 cm^{-1} and 1574 cm^{-1} due to ring mode stretch and the C=C backbone stretching of PPy. The nanocomposite decrease in the I_D / I_G value showed that the disorder graphene is less in the composite and manifested in the G-band peak from 1572 cm^{-1} to 1566 cm^{-1}.

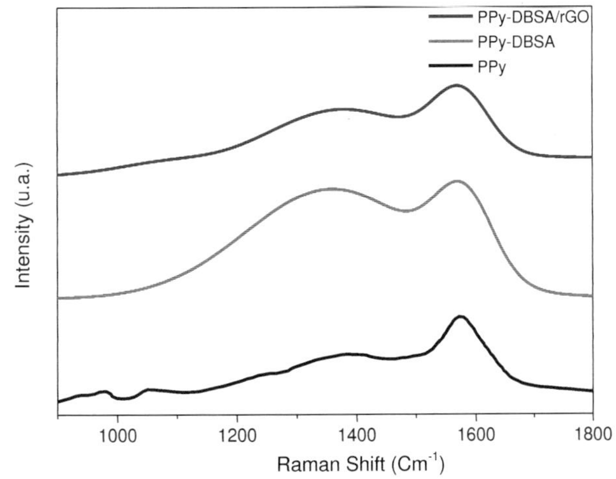

Figure 3. Raman spectra for the PPy, PPy-DBSA, and PPy-DBSA/rG O

3.2 SEM Characterization

Figure 4 demonstrates PPy-DBSA/rGO nanocomposite morphology. On the surface of reduced graphene oxide nanosheets, PPy has a wrinkled shape generating a mesh structure. This morphological characteristic provides a broad adsorption surface contact and enhances the adsorption of NO2 molecules.

Figure 4. SEM images of PPy-DBSA/rGO films grown on surface ITO/PET.

3.3 Reproductibility and response time of thin films of PPy-DBSA/rGO

The reproducibility test for the PPy-DBSA/rGO thin-film based NO_2 sensor was evaluated by plotting the response curve in dynamic mode after exposure of the film to different concentrations of nitrogen oxide ranging from 1ppm up to 20ppm, as shown in Fig. 5. The response and recovery time showed in fig. 6 were estimated 30.25 s and 75 s respectively.

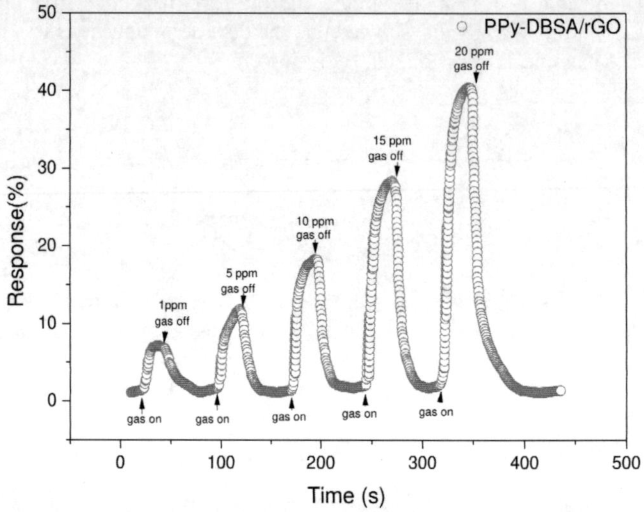

Figure 5. Reproducibility test of thin film Sensor PPy-DBSA/rGO

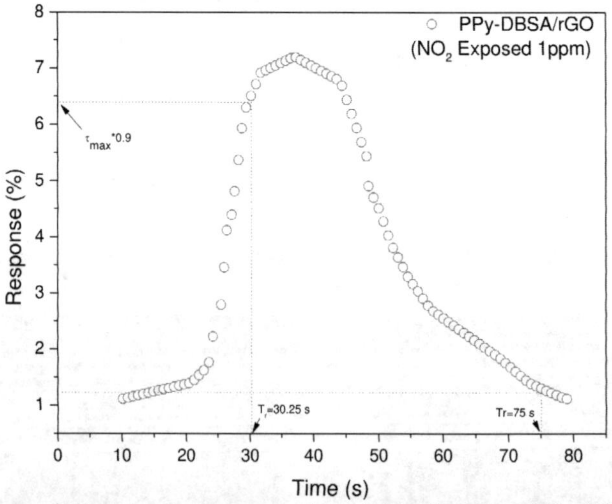

Figure 6. Response and recovery time of thin film PPy-DBSA/rGO

3.4 Effect of relative humidity on the sensor response

the injection of 2-ppm of NO2 gas was carried out under a humid atmosphere to scrutinize the effect of humidity on the sensor performance. A hygrometer (Omega type HH314A) was used to indicate the relative humidity (RH) generated by injecting water vapor in the test chamber at rates ranging from 20% to 80%, as illustrated in Fig. 6. Under a humidity rate higher than 50%, a substantial humidity impact on sensor efficiency was detected. This result may be due to the inhibition of the active sites by -OH groups, when they are bridged with -COOH entities in surfactant agent through a hydrogen bond.

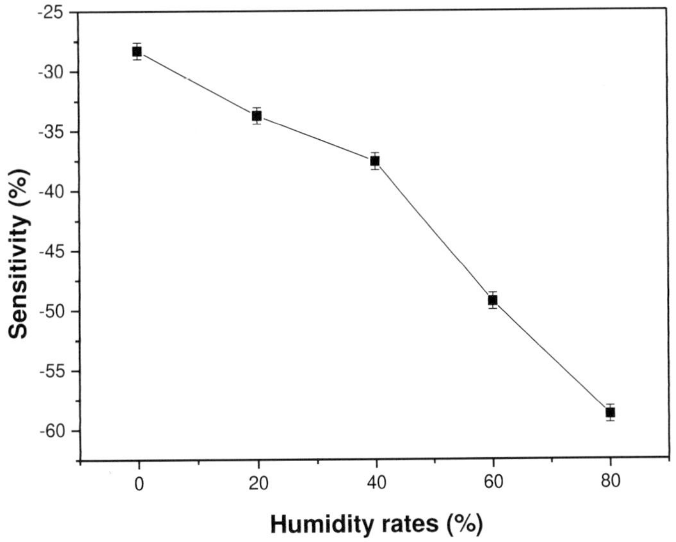

Figure 7. PPy-DBSA/rGO sensor responses to NO_2 vapors at different humidity atmosphere rates.

4. Conclusion

A newly flexible nanocomposite-based nitrogen dioxide chemisensor was developed from polypyrrole filled with graphene doped with DBSA using electro polymerization technique. The addition of DBSA and reduce graphene oxide enhanced the electrical conductivity of as-prepared nanocomposite. Results showed that the produced PPy-DBSA/rGO sensor demonstrated improved NO_2 sensing properties including response time (30.25 s), the detection limit (2ppm), and reproducibility. Furthermore, Results showed that the rGO exfoliated by n-dodecylbenzene sulfonic acid improved both the sensitivity and the reactivity of our produced nanocomposite toward NO_2. Higher NO_2 response was detected for this nanocomposite in comparison to several studied gases such as NO, NH3, H_2S, and CO_2. Furthermore, its sensitivity remained considerable under varied temperature and humidity, indicating its good function ability under harsh environment film. confirmed that this sensor displayed a resistor–like type.

References

[1] Parthasarathy Srinivasan, Madeshwari Ezhilan, Arockia Jayalatha Kulandaisamy, K. Jayanth Babu, John Bosco and Balaguru Rayappan 2019 J. Mater. Sci: Mater Electron 30, 15825–15847
[2] T. Boningari, P. G. Smirniotis. Curr. Opin. Chem. Eng. 13, 133-141 (2016)
[3] Jian Zhang, Ziyu Qin, Dawen Zeng, Changsheng Xie. Phys. Chem. Chem. Phys. 19, 6313-6329 (2017)
[4] A. T. Mane, S. D. Sartale, V. B. Patil. J Mater Sci: Mater Electron 26, 8497–8506 (2015)
[5] M. Das, S. Roy. Mater. Sci. Semicond. Process. 121, 105332 (2021)
[6] S. Geetha, D. C. Trivedi. J Mater Sci: Mater Electron 16, 329–333 (2005)
[7] Shruti Peshoria and Anudeep Kumar Narula. J Mater Sci: Mater Electron. 28, 18348–18356 (2017)
[8] R. Bissessur, P. K. Y. Liu, and S. F. Scully. Synth. Met. 156, 1023-1027 (2006)
[9] S.G. Pawar, S.L. Patil, M.A. Chougule, B.T. Raut, S.A. Pawar, R.N. Mulik and V.B. Pati l 2012 J. Mater. Sci. Mater. Electron. 23 273–279
[10] D. Guettiche, A. Mekki, B. Lilia, T. Fatma-Zohra and A. Boudjellal. J. Mater. Sci.: Mater. Electron. (2021) ,10662–10677
[11] F. Merdj, A. Mekki, D. Guettiche, B. Mettai, Z. B. D. Sayah, Z. Safidine, A. Abdi and R. Mahmoud. Macromol. Res. (2018), 26 511–520

The efficiency of steam production by the Parabolic Trough Collector PTC with the use of nanofluids in the region of Ouargla

Intissar Achouri[1], Mouhamed Elbar Soudani[1], and Tlili Salah[2]

[1] Laboratoire de Développement des Energies Nouvelles et Renouvelables dans les Zones Arides et Sahariennes, Faculté des Mathématiques et des Sciences de la Matière, Université Kasdi Merbah Ouargla, Ouargla 30000, Algeria
[2] Department physic, Faculté des Mathématiques et des Sciences de la Matière, Université Kasdi Merbah Ouargla, Ouargla 30000, Algeria.

intissarachouri.05@gmail.com

Abstract. Concentrated solar power plants (CSP) contribute to global production (at present) with a capacity of 400 MW, and by 2020 they will reach approximately 20 GW, then nearly 800 GW by 2050, This will prevent the emission of 32 million tons of CO2 annually in 2020, and rise to 1.2 billion tons in 2050, according to the International Greenpeace "Solar Thermal Electricity" 2016 report.
Among all the concentrated solar power (CSP) technology available to date, Parabolic Trough Collector (PTC) is the most promising, cost-effective, and efficient solution to generating electrical power, as PTC plants contribute in terms of global production capacity by 73.58% of the overall capacity of concentrated solar power plants (CSP).
PTC stations in the production of electricity depend on the generation of hot and pressurized steam that rotates the turbines and to increase the effectiveness of PTC in the production of steam, we use in this study nanofluids by adding copper nanomaterials in different proportions to improve the Thermal efficiency of PTC. We also studied the effect of the width of the PTC slot on the fluid temperature. And from it on the amount of steam produced.
The results of the study showed that the Thermal efficiency increases with the increase in the ratio of copper nanomaterials in the water, as the temperature of outlet water reaches 98°C, for the ratio of nanomaterials, 20%, in order to water flow 0 .01 Kg/s and display the aperture3.5 m.
Keywords: Parabolic Trough Collector, nanofluid, Thermal efficiency, steam.

1. Introduction

With the excessive doubling of global energy demand since after the industrial revolution and the consequent pollution of fossil fuel combustion and climate change, which leads to threatening the ecological balance of the environment on one hand and the depletion of the world's reserves of fossil fuels on the other hand, the world is in a race towards Exploitation and development of renewable energy sources (solar radiation, wind and biomass, Earth's gravity, tides and the heat of the Earth's interior) which are available and environmentally friendly.
The sun is one of the most important energy sources, as the energy flux density emitted from it is 63 MW / m2 in all directions, while the Earth receives from it 1.7 1014 KW, and the energy flux density on the surface of the earth is within the limits of 1KW/ m^2 [2,1].
The concentrated solar plant (CSP) technology is reasonably cost-effective and efficient. Of all the solar concentrated power (CSP) technologies available to date, the Parabolic Drum accumulators (PTC) are the most promising and cost-effective solution for generating electricity [3]. PTC plants contribute in

terms of global production capacity with 73.58% of the capacity of concentrated solar power plants (CSP) in general [4].
Choi was the first to use the term nanofluids since 1995 when he suggested the possibility of multiplying the heat transfer coefficients by convection using these fluids. In addition to researchers in Japan and Germany publishing articles related to similar fluids [5], Heris et al have experimented with Al2O3 and CuO nanoparticles suspended in water. They found that heat transfer could increase by about 40% while improving thermal conductivity did not exceed 15% [6].
Otanicar et al. In 2010 studied the experimentally and numerically effect of various nanofluids on the performance of small-scale direct absorption solar collectors. Their results showed that suspending a slight amount (<0.5%) of the nanoparticles significantly improves the efficiency. However, for a volume fraction above 0.5%, the efficiency remains constant and even begins to decrease as the fraction size increases. They also found that the efficiency increases with decreasing the size of the nanoparticles [7].
In this study, nanofluids are used by adding copper nanomaterials in different proportions to improve the thermal yield of PTC. We also studied the effect of the width of the PTC opening on the fluid temperature and from it on the amount of ateam produced that can be exploited in various industrial and home applications.

2. Theoretical study

In order to know the effectiveness of vapor production by the PTC parabolic center using nanofluids in different proportions and after calculating and determining the direct solar radiation. As well as the properties of the nanofluid, where the physical properties of the nanofluids can be expressed in terms of the physical properties of the base solution and the nanomaterials added to it according to the following relationships: [8].
The density of the nanofluid is written as follows:

$$\rho_{nf} = \rho_{bf}.(1 - \phi) + \rho_{np}.\phi_f \qquad (01)$$

The heat capacity of nanofluids is written as follows:

$$C_{p.nf} = \frac{\rho_{bf}.(1-\phi)}{\rho_{nf}}.C_{p.bf} + \frac{\rho_{np}\phi}{\rho_{nf}}.C_{p.np} \qquad (02)$$

The conductivity of a nanofluid can be expressed as:

$$K_{nf} = K_{bf}\frac{K_{np}+2K_{bf}+2(K_{np}-K_{bf})\phi}{K_{np}+2K_{bf}-(K_{np}-K_{bf})\phi} \qquad (03)$$

Nanofluid viscosity is written as follows:

$$\mu_{nf} = \mu_{bf}(1 + 2.5\phi + 6.2\phi^2) \qquad (04)$$

3. PTC model thermal equilibrium

Thermal equilibrium equations for all surfaces of the components of the PTC, across a cross-section of the system, based on the first law of thermodynamics.

*For the fluid "**f**":*

$$m_f.C_{p.f}\frac{dT_f}{dt} = h_{conv\,(r-f)}(T_r - T_f)A_{r.int} \qquad (05)$$

*For the absorbent tube "**r**":*

$$\rho_r A_r C_{pr}\frac{dT_r}{dt} = IC_g\tau_e\alpha_r\rho^{\circ}\pi Dr_{ext} + \left(h_{conv\,(r-e)} + h_{rad\,(r-e)}\right)(T_e - T_r)\pi Dr_{ext} + h_{conv\,(r-f)}(T_f - T_r)\pi Dr_{int} \quad (06)$$

For the glass envelope "e":

$$\rho_e A_e C_{pe}\frac{dT_e}{dt} = IC_g\alpha_e\rho^{\circ}\pi De_{ext} + \left(h_{conv\,(e-r)} + h_{rad\,(e-r)}\right)(T_r - T_e)\pi De_{int} + \left[h_{conv\,(e-amb)}(T_{amb} - T_e) + h_{rad\,(e-sky)}(T_{sky} - T_e)\right].\pi.De_{ext} \quad (07)$$

4. Results and Discussion

4.1. Characteristics of the study area and its day

The region of Ouargla is located at longitude $\lambda = 5.24°$, latitude $\varphi = 31.57°$, and H = 141 m above sea level. This excellent location makes it a good place to exploit and value the enormous solar energy, in several fields: electrical using solar cells, and thermal directing and focusing on solar collectors such as Fresnel lens and parabolic collector used in our study. In this region, the solar radiation flux reaches 2650 Kwh / an / m2, with about 3500 hours of insolation per year [9].

In our study, we chose the coldest day of the year, which we find out by calculating the air temperature or The ambient during the day, and at every moment of solar time, using the following relationship [10]:

$$T_{amb} = \left(\frac{T_{amax}+T_{amin}}{2}\right)\left[\sin\left(\frac{(t-8)\pi}{12}\right) + 1\right] \quad (08)$$

By measuring the daily temperatures at the meteorological stations located in the region of Ouargla over a period of ten years (1996-2006), and after calculating the monthly average of the highest and lowest temperatures, it was found that January is the coldest (5.1-18.6 °C) [11].

It was also found by calculation and through the previous relationship that the lowest temperature in this region is on the twentieth day of this month, as T_{amb} is estimated =19.89 °C, which appears after one o'clock in the afternoon by about 52 minutes and 12 seconds, which is evident in figure (1) below. Which shows the evolution of this temperature with solar time.

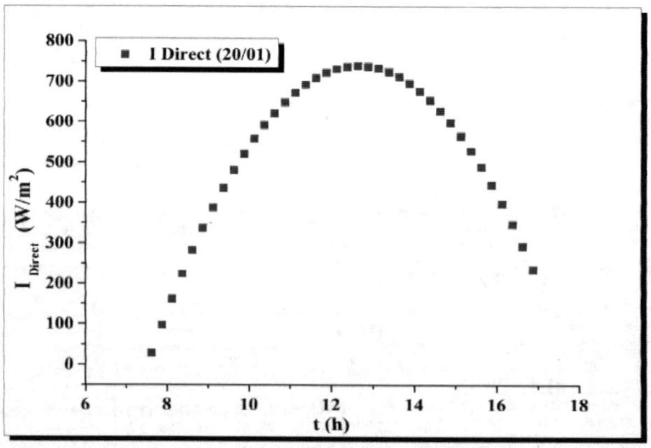

Figure 1. presents the change in the ambient temperature on the 20th of January in the Ouargla region with the change of solar time.

It can also be calculated by obtaining the change in the intensity of direct solar radiation with the solar time on this day, represented in the following figure (2):

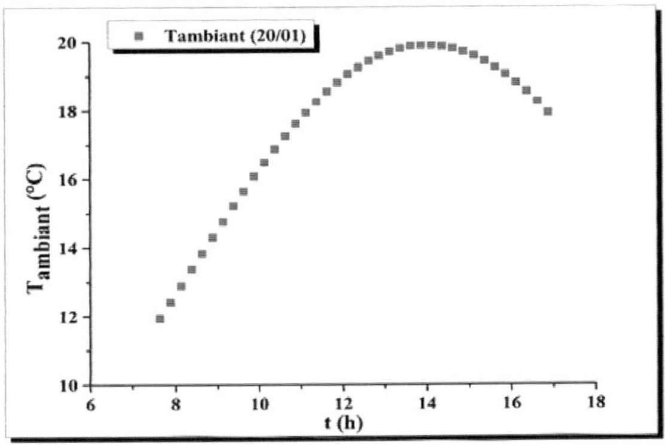

Figure 2. represents the change in the intensity of direct solar radiation theoretically calculated on the 20th of January in Ouargla region with the change of solar time.

From the figure, it is noticed that there is an increase in the intensity of the solar radiation until midday and 37 minutes and 8 seconds, where it reaches its maximum value estimated at about 740w/ m^2, and then retracts until it ceases to exist at sunset, as the duration of insolation lasted nine hours and a quarter.

4.2. Study of the fluid outlet temperature

In this study, the parabolic collector supposed to be used as a length of L=5 m and a width that varies between 3 to 3.5 m, passing the value of 3.25 m, and with a nanofluid formed from nanoparticles of copper metal placed in water, the ratio of which changes from zero to up to 20 %, Also by 10%, with a constant flow through the collector tube of 0.01kg/s.

Through Figures (3 and 4), we notice:

The temperature of the fluid outlet increased with time in a linearly, starting from ten in the morning until one midday until it stabilized at a maximum value before sunset, until it stabilized at a maximum value before sunset, as this temperature exceeds the ambient temperature twice as low as possible (see table). This table contains the values of the maximum fluid outlet temperature and the time of its emergence for each collector, along with the change of the ratio of nanoparticles in the water.

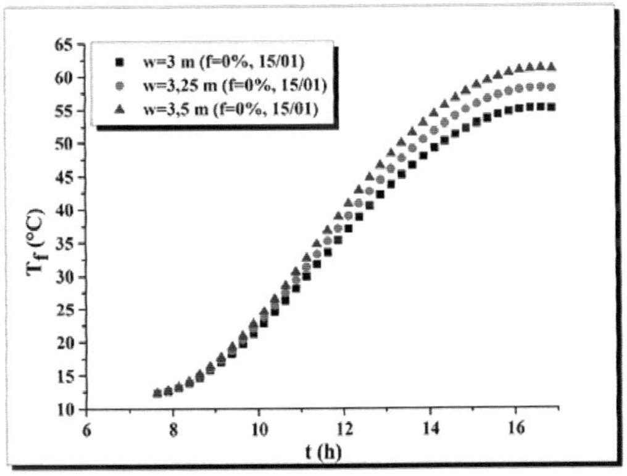

Figure 3. shows the change in the outlet temperature of the water (the nanofluid ratio is null) with the change of the center width and the change of the solar time.

Through both Figures (3) and Table (1), it is clear that the outlet temperature of the fluid increases with the increase in the width of the collector, starting from ten in the morning.
As for Fig (4) and Table (1) also, the increase in the outlet temperature of the fluid can be clearly seen with the increase in the presence of nanoparticles in the water, starting from nine in the morning. The highest temperature is reached in the case of the collector with a width of 3.5 m and for the fluid of the particle percentage present in one hundred to 20%.

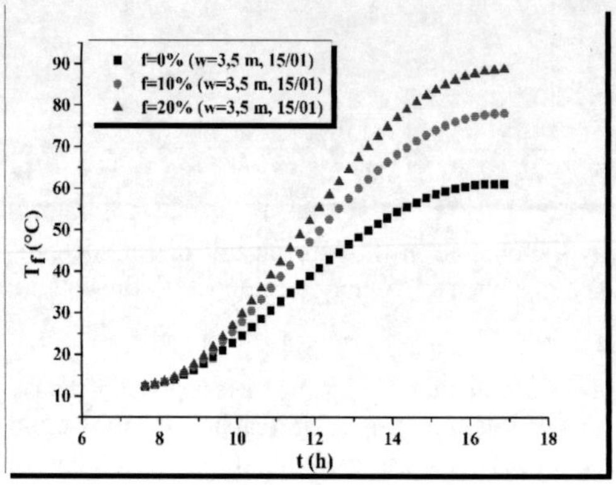

Figure 4. shows the change in the outlet temperature of a nanofluid with a change in the presence of the copper nanoparticles and the solar time.

Table 1. presents the maximum outlet temperatures of the fluid, the time of its emergence, and the temperature stored in the fluid from each collector, with the change in the proportion of nanoparticles in the water.

PTC dimension	Proportion (%)	Température (Time)	T stored fluid
	0	55,19 °C (16.63 h)	36.08 °C
5 m . 3 m	10	58.21 °C (16.63 h)	37.83 °C
	20	61.11 °C (16.63 h)	39.53 °C
	0	70.35 °C (16.88 h)	43.68 °C
5 m . 3,25 m	10	74.30 °C (16.88 h)	46.06 °C
	20	78.01 °C (16.88 h)	48.34 °C
	0	80.11 °C (16.88 h)	48.44 °C
5 m . 3,5 m	10	84.60 °C (16.88 h)	51.22 °C
	20	88.73 °C (16.88 h)	53.87 °C

4.3. Study the temperature of the stored fluid
Through Figure (5), it can be observed that the temperature of the copper nanofluid is changed by 20% after leaving the collector with a width of 3.5 m, on the basis that this storage is in a thermally insulated tank. As it appears that the temperature change here is similar to the change in the fluid outlet

temperature in general, as it reaches 53.87 ° C at sunset, which is twice the highest temperature of the ambient, with a stored value in the mass of fluid estimated at 342 kg. This is a mass suitable for domestic use with a limit of water Despite its weakness.

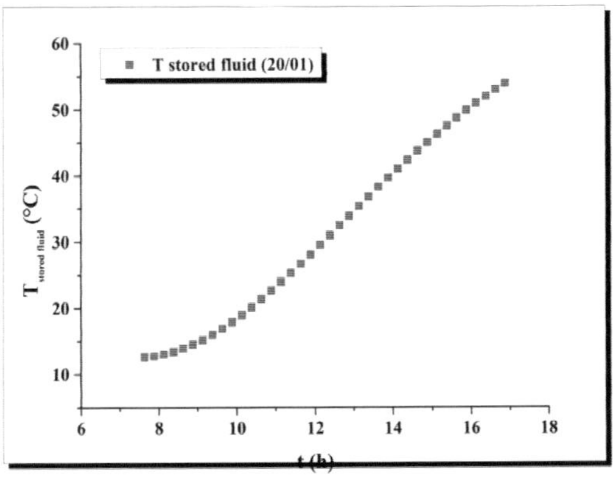

Figure. 5. shows the temperature change of the stored nanofluid with the change in solar time.

In Table .1. the values of the temperature of the stored fluid are recorded for the various cases in which the study was conducted, and the increase in the temperature of the stored fluid is confirmed by an increase in both the width of the collector and the percentage of nanoparticles in the water.

5. Conclusion

This work focused on changing the width of the 5 m long PTC collector and the ratio of copper nanoparticles placed in 0.01 kg / s water flow, leading to the following conclusion:
1- The two degrees of fluid outlet temperature and the temperature stored in it increase with the increase in the width and the ratio together, as the two increases are the same (figure).
2- The fluid outlet temperature can be doubled from the ambient temperature to more than four times, while the temperature stored in it can be doubled twice.
3- The amount of heat storing material is small, but it can be considered sufficient for personal use within the small dimensions of the collector.

References

[1] Kreith F, Kreider JF. Principles of solar engineering. New York: McGraw-Hill; (1978).
[2] M. Romero, D. Martìnez, and E. Zarza, "Terrestrial solar thermal power plants: on the verge of commercialization," in Proceedings of the 4th international conference on solar power from space, (2004).
[3] O. Behar, A. Khellaf and K. Mohammedi, A novel parabolic trough solar collector model - Validation with experimental data and comparison to Engineering equation solver (EES), Energy Conversion and Management, 106 (2015) 268-281.
[4] D. A. Baharoon, H. A. Rahman, W. Z. W. Omar and S. O.Fadhl, Historical development of concentrating solar power technologies to generate clean electricity efficiently - a review, Renew Sustain Energy, Rev., 41 (2015) 996-1027.
[5] K. Das, U.S. Choi, Wenhua Yu, T. Pradeep, Nano-fluids science and technology,John Wiley & Sons.
[6] S. Z. Heris, S. Gh. Etemad, M. N. Esfahany, Experimental investigation of oxide nano-fluids laminar flow convective heat transfer, Int Commun Heat Mass Transfer, vol. 33, pp. 529-535, (2006).
[7] M. E. Soudani, K. E. Aiadi, D. Bechki, "Water heating by Parabolic Trough Collector with storage in the Ouargla region of Algerian Sahara", Materials Today: Proceedings, 24, 137–

139, (2020).
[8] N. Basbous, M. Taqi, N. Belouaggadia NUMERICAL STUDY OF A PARABOLIC TROUGH COLLECTOR USING A NANOFLUID 2015 Asian Journal of Current Engineering and Maths 4:3 May - June (2015) 40 – 44
[9] Marif. Y, Contribution à l'amelioration du rendement d'un distillateur a film capillaire multi étage. Mémoire de magister en physique énergétique. U. K. M. Ouargla; (2008).
[10] Egbo G., Sintali I. S., and Dandakouta H., "Analysis of Rim Angle Effect On the Geometric Dimensions of Solar Parabolic-Trough Collector", Int. Jor. P. App. Scs., 2 (2008) No.3, 11-20.
[11] M. E. Soudani, K. E. Aiadi, D. Bechki and S. Chihi, Experimental and theoretical study of Parabolic trough collector (PTC) with a flat glass cover in the region of algerian sahara (Ouargla), Journal of Mechanical Science and Technology 31 (8) (2017) 4003~4009.

IOP Conf. Series: Materials Science and Engineering 1204 (2021) 012006 doi:10.1088/1757-899X/1204/1/012006

Wastewater treatment by adsorption process on mineral actived carbon: modeling and prediction using an intelligent artificial approach

Sediri Meriem[1], Hanini Salah[1]

[1] Biomaterials and Transport Phenomena Laboratory (LBMPT), University of Médéa, Ain D'Heb 26000, Médéa, Algeria

sedirim2@gmail.com

Abstract. Currently there are several wastewater treatments processes, and several adsorbent materials consist of separating and purifying the various industrial effluents. In this work an artificial neural network (ANN) was developed to describe the dynamic adsorption of sodium decanesulfonate using actived carbon obtained by the calcination of mineral biomass under different conditions. Three inputs (time, mass of adsorbent and fixed bed height) were used in the input layer, three neurons in the hidden layer and one in the output layer for the reduced concentration. The Levenberg Marquardt back-propagation algorithm was applied. The tangent sigmoid and linear transfer functions are used for the hidden layer and the output layer respectively. The results showed a correlation coefficient $R^2 = 0.9965$ with root mean squared error RMSE = 0.0276. An interpolation and an extrapolation stage are made to test the accuracy of the network. The results showed a high correlation coefficient $R^2 = 0.9969$ and 0.984 respectively for the interpolation and the extrapolation. These results show the robustness and the high capacity of ANN to describe the dynamic adsorption of sodium decanesulfonate onto actived carbon.
Keywords. Water treatment, fixed bed adsorption, Mineral biomass, Modelling, Artificial Neural Network,

1.Introduction

Water bodies are the most common natural resources that have been contaminated as a result of different human activities [1]. These wastewaters containing toxic compound coming from different industries had a harmful impact on the atmosphere and their reject has dangerous results can be lethal to the aquatic life and affecting directly on the ecosystem equilibrium for that they must be treated properly before their discharge to the environment [2,3].

Adsorption has been extensively studied by researchers worldwide as an efficient and economically sustainable technology for the treatment of effluents [4].
Fixed bed adsorption or dynamic adsorption is widely used for purifying liquid mixtures, especially industrial waste effluents. This technical is more appropriate due to its low cost and high efficiency [5,6].
Artificial neural networks (ANN) are computational models as set of algorithms; they resemble the human brain [7]. They are now used as a very powerful tool to model nonlinear systems and they are considered more reliable than other mathematical models because they have a high parallelism and a high robustness [8, 9].

The ANN architecture containing a number of hidden layers, input layer (independent variables), and output layer. Each of these layers consists of a number of interconnected processing units called neurons. These neurons are connected between them by links or weights [10] .

Several works have used ANN as a black box to predict the dynamic adsorption phenomena.

(Kittisupakorn P et al.) [11] developed an artificial neural network to predictthe dynamic adsorption of Hydrochloric Acid onto Ion Exchange Resin. (Celekli A et al.) [12] studied the removal modeling of Lanaset Red G on Chara contraria by ANN. (Sediri M el al.) [13] in their work, they developed an ANN to predict the fixed bed adsorption of p-nitrophenol onto NDA-100 resin.

The aim of this work is to develop an artificial neural network (ANN) to predict the dynamic adsorption of sodium decansulfonate onto actived carbon under different conditions. The accuracy of the optimal network obtained was tested by studying the interpolation and the extrapolation performance.

2.Materials and methods

2.1. Materials

The system (adsorbate /adsorbent) studied in this work was the Sodium decanesulfonate/actived carbon. Experimental data points were taken from the article of Sahel and Ferrandon-Dusart [14], and arranged as matrix containing 156 experimental data points. The actived carbon properties are shown in Table 1.

Table 1. Properties of actived carbon [14]

Characteristic	value
Area surface (m^2/g)	1100
Particle diameter(mm)	1
Apparent density (kg/m^3)	535

2.2. Methodology

The feed-forward back propagation neural network (FFNN) with Levenberg Marquardt back-propagation Learning algorithm is one of the most famous neural network paradigms used in the modeling of non linear systems which has been used in our study. For the training/testing of the FFNN-LMBP, 156 experimental data points were collected and arranged as matrix of dimension of (156, 4).

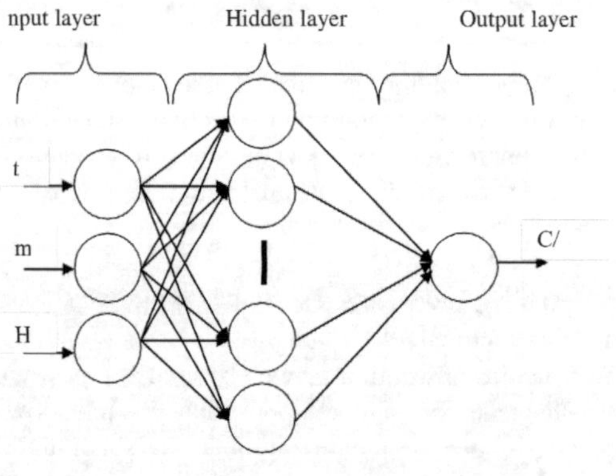

Figure 1. Topology of neural network

Three (03) parameters (time (t), mass of adsorbent (m) and fixed bed height (H)) were used as inputs in the input layer of the ANN and the reduced concentration (C/C_0) selected as the output parameter in the output layer. The ANN architecture is shown in Fig.1.

The normalization of the experimental data in the interval [-1, 1] was achieved using the mapminmax function given by the following equation [15]:

$$X_{norm} = Y_{min} \left(\frac{X_i - X_{min}}{X_{max} - X_{min}}\right)(Y_{max} - Y_{min}) \quad (1)$$

Where X_i is the input or output variable X, X_{min} and X_{max} are the minimum and maximum values of variable X, $Y_{max} = +1$, $Y_{min} = -1$.

The splitting of the experimental data was also achieved into three phases: training, validation and testing phase. For the hidden layer the number of neurons was changed from 1 to 10. The tangent sigmoid (tansig) and linear (purelin) activation function used at hidden and output layer respectively.

The performances of the neural network were calculated using the root mean square error (RMSE) and correlation coefficient (R^2) given by the following equations [13,16]:

$$R^2 = 1 - \frac{\sum_q^p \left[\left(\frac{C}{C_0}\right)_q^{exp} - \left(\frac{C}{C_0}\right)_q^{predict}\right]^2}{\sum_q^p \left[\left(\frac{C}{C_0}\right)_q^{exp} - \left(\frac{C}{C_0}\right)_n\right]^2} \quad (2)$$

$$RMSE = \sqrt{\frac{1}{p}\sum_{q=1}^{p} \left[\left(\frac{C}{C_0}\right)_q^{exp} - \left(\frac{C}{C_0}\right)_q^{predict}\right]^2} \quad (3)$$

Where, q is the number of data points, $\left(\frac{C}{C_0}\right)^{exp}$ is the experimental reduced concentration and $\left(\frac{C}{C_0}\right)^{predict}$ is the predicted reduced concentration by the ANN model.

3. Results and discussion

3.1. ANN developed model

In order to determine the optimal structure of ANN, the number of neurons was changed at hidden layer from 1 to 10 using the tansig activation function. Each topology was repeated three times. According to the values of correlation coefficient (R^2) and root mean square error (RMSE), three neurons are obtained at hidden layer corresponds to the agreement vector [R^2 , RMSE] approaching to the ideal [R^2=1, RMSE=0]. Fig .2 shows the comparison between experimental and predicted data by ANN model. This plot has $R^2 = 0.9965$ and RMSE= 0.0276 provide the reliability of the developed ANN model.

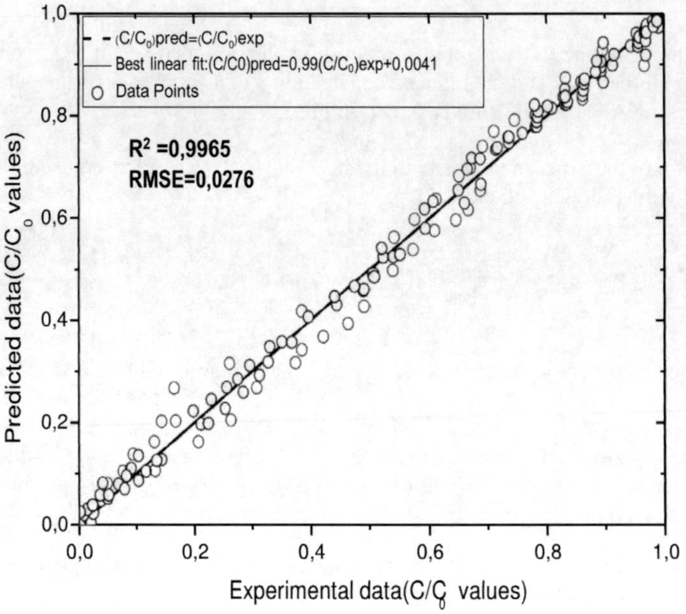

Figure 2. Comparison between calculated values by ANN and experimental values

The breakthrough curves of sodium decanesulfonate as a function of time under different conditionswere depicted in Fig .3. The experimental data points are very approaching to those obtained by the ANN, which shows the robustness of neural network model.

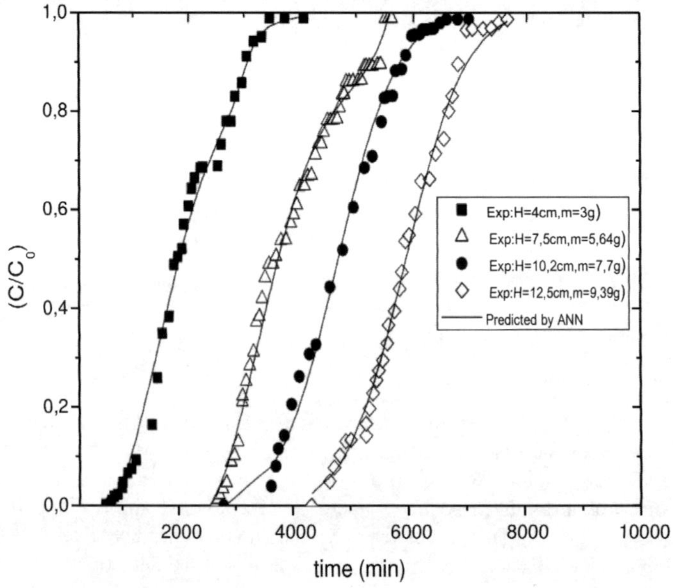

Figure 3. Breakthrough curve of sodium decanesulfonate onto actived carbon, C_0=20mg/l, flow rate Q=0.435l/h

3.2. Interpolation performance

An interpolation was studied to check the precision of the ANN model developed. A data base was used containing a set of data points situated at the intermediate of the experimental points of a breakthrough curve of sodium decanesulfonate / actived carbon system. The results of interpolation performances in terms of error (RMSE) and agreement vector values [α= (slope), β (y intercept), R^2 (correlation coefficient] are presented in Table 2. The plot of breakthrough curve of interpolated experimental data and predicted data is shown in Fig.4. These results showed the robustness of ANN developed model.

Table 2. Interpolation results in terms of error (RMSE) and vector [α, β, R^2]

Interpolation performances			
RMSE	R^2	*α*	*B*
0.0274	0.9969	0.98	0.01

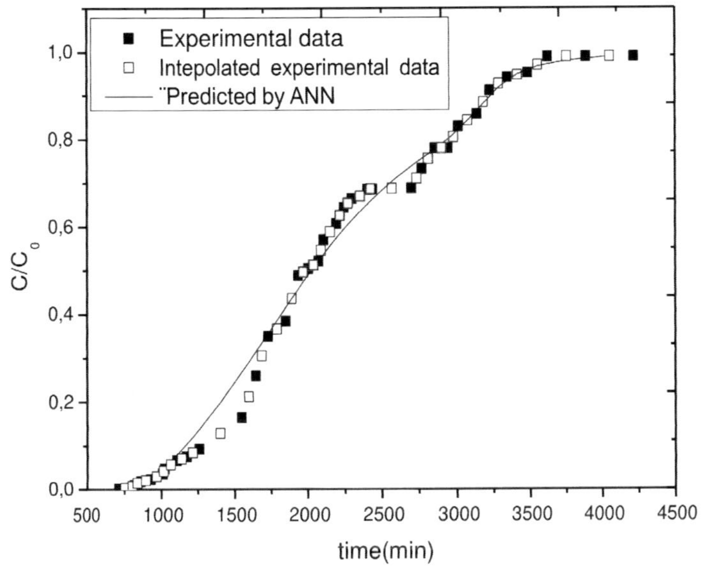

Figure 4. Simulation results of interpolation: C_0=20mg/l, flow rate Q=0.435 l/h, H=4cm, m=3g.

3.3. Extrapolation performance

An extrapolation was also studied to check the accuracy of the ANN model developed. For that we have testing an experimental data point never exploited during the learning and the testing phases of ANN previously developed. Results of extrapolation performances were showed in Table 3. Fig .5 showed the breakthrough curve of extrapolated experimental data with those obtained by ANN. The results provide that the ANN model predicts the experimental data very well.

Table 3. Extrapolation results in terms of error (RMSE) and vector [α, β, R^2].

Extrapolation performances			
RMSE	**R^2**	***α***	***Β***
0.0745	0.984	0.91	0.019

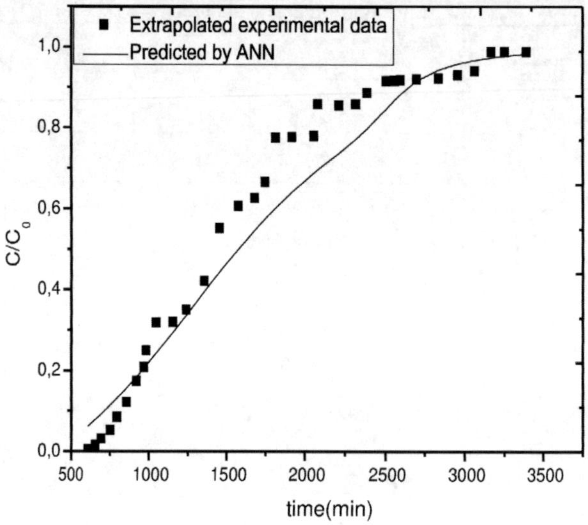

Figure 5. Simulation results of extrapolation: C_0=20mg/l, flow rate Q=0.435 l/h, H=3.10cm, m=2.33g.

4. Conclusion

An ANN was developed in this work to predict the dynamic adsorption of sodium decanesulfonate using actived carbon as adsorbent. For this purpose, three neurons were used in the input layer, three in the hidden layer with tansig activation function and one neuron is used in the output layer with purelin activation function. Levenburg-Marquardt learning algorithm was applied. The results showed a high correlation coefficient of order of 0.9965 with root means square error close to 0.

The high correlation coefficients obtained of 0.9969 and 0.984 values for the interpolation and for the extrapolation stage respectively showed the accuracy and the robustness of the neural network developed model.

References

[1] Cavus S and Gurdag G 2008 Competitive heavy metal removal by poly (2-acrylamido-2methyl-1-propane sulfonic acid-co-itaconic acid. *J .Polymer. Advan. Technol.* pp 1209 –1217.

[2] Amegrissi F, Talbi M, El Kouali M, Dahbi L, Ainane T and Maghri I 2012. *Mater. Environ. Sc. pp* 744-753.

[3] Ainane T, Abourriche A, Kabbaj M, Elkouali M, Bennamara A, Charrouf M and Talbi M 2014 *J.Chem. Pharm. Res.* pp 599-606.

[4] Mittal A, Kurup L and Gupta VK 2005 Use of waste materials—bottom ash and de-oiled soya, as potential adsorbents for the removal of Amaranth from aqueous solutions. *J. Hazard. Mater*. pp 171–178.

[5] Xu Z, CAI J-g and PAN, B- C 2013 Mathematically modeling fixed-bed adsorption in aqueoussystems . *J Zhejiang Univ-Sci A (Appl Phys& Eng).* pp 155-176.

[6] Aribike D S and Olafadehan O.A 2008 Modeling of Fixed Bed Adsorption of Phenols on Granular Activated Carbon. *Theoretical Foundations of Chemical Engineering*. pp 257–263.
[7] Haykin S 1994 Neural Networks: A Comprehensive Foundation. *NY: Macmillan*. p.2.
[8] Basheer I A and Hajmeer M 2000 Artificial neural networks: Fundamentals, computing, design,and Application. J . Microbiol. Methods .pp 3-31.
[9] Plumb AP, Rowe RC, York P and Brown M 2005 Optimisation of the predictive ability of artificial neural network (ANN) models: A comparison of three ANN programs and four classes oftraining algorithm. *Eur J Pharm Sci*. p 395.
[10] Aghav RM, Kumar S and Mukherjee SN 2011 Artificial neural network modelling in competitive adsorption of phenol and resorcinol from water environment using some carbonaceousadsorbents. *J. Hazard. Mater*. pp 67–77.
[11] Kittisupakorn P, Tangteerasunun P and Thitiyasook P 2005 Dynamic Neural Network Modelingfor Hydrochloric Acid Recovery Process. *Korean J. Chem. Eng*. pp 813-821.
[12] Celekli Aand Geyik F 2011 Artificial neural network (ANN) approach for modelling of removalof Lanaset Red G on Chara contraria. *Bioresour. Technol*. pp 5634–5638.
[13] Sediri M, Hanini S, Cherifi H, Laidi M and Abbas Turki S 2017 Dynamic Adsorption Modellingof P-nitrophenol in Aquous Solution Using Artificial Neural Network. *JMater .Environ. Sci*. pp 2282-2287.
[14] Sahel M and Ferrandon-Dusart O 1993 Adsorption dynamique en phase liquide sur charbon actif : comparaison et simplification de différents modèles. *Rev Sci Eau*. pp 63-80.
[15] Yacef R, Mellit A, Belaid Sand Sen Z 2014 New combined models for estimating daily global solar radiation from measured air temperature in semi-arid climates: Application in Ghardaïa, Algeria.*Energ Convers Manage*. pp 606–615.
[16] Cavas L, Karabay Z, Alyuruk H, Doğan Hand Demir GK 2011 Thomas and artificial neural network models for the fixed-bed adsorption of methylene blue by a beach waste *Posidonia* oceanica(L.) dead leaves. *Chem Eng J*. pp557 - 562 .

Elaboration of a piezoelectric force sensor for medical application

Zakia Chelli[1,2], Hakim Achour[1,2], Malika Saidi[1,3], Mourad Laghrouche[2,4], Ahcene Chaouchi[1,3]

[1] Department of Chemistry, Mouloud Mammeri University, Tizi-Ouzou, Algeria

[2] L.A.M.P.A Laboratory, Department of Electronics, Mouloud Mammeri University, Tizi-Ouzou, Algeria;

[3] L.C.A.G.C Laboratory, Chemical Department, Mouloud Mammeri University, Tizi-Ouzou, Algeria

[4] Department of Electronic, Mouloud Mammeri University, Tizi-Ouzou, Algeria

zakiach.1704@gmail.com

Abstract. In this work, a piezoelectric force sensor was fabricated based on Lead-Free ferroelectric particles $(Na_{0.535}K_{0.48})_{0.966}Li_{0.058}Nb_{0.9}Ta_{0.1}O_3$ (NKLNT) synthesized by solid-state route. A piezoelectric characterization by FTIR spectroscopy and dielectric characterization was performed.

The manufactured NKLNT sensor shows excellent sensitivity and response to external mechanical forces, which implies that this sensor presents a promising technology for different applications.

Keywords. Piezoelectric force sensor; Ferroelectric particules; solid-state route; NKLNT

1. Introduction

Piezoelectric materials capable of converting mechanical energy into electrical energy (generator or sensor mode) or vice versa (actuator mode) [1]. These materials are widely used in many technological applications and are incorporated in various electrical and electronic equipment such as actuators, sensors, transducers, etc. Piezoelectric ceramics such as lead Zirconate Titanate (PZT) [2] have very good dielectric and piezoelectric properties (high dielectric constant), but they have the disadvantage of containing lead, which has a polluting and therefore harmful impact on the planet. Therefore they must be replaced by other lead-free compositions such as $(Na_{0.535}K_{0.48})_{0.966}Li_{0.058}Nb_{0.9}Ta_{0.1}O_3$ particles (NKLNT) [3].

These materials are widely used in many technological applications and are incorporated in various electrical and electronic equipment such as actuators, sensors, transducers, etc. Piezoelectric sensors have several advantages, such as high sensitivity, wide frequency range, high durability, and fast response. Thus, these sensors can be widely used in several fields: Aeronautics [4], industry [5], energy recovery devices [6-8], and also in the medical field [9]. In the literature, there are different synthesis methods for the elaboration of piezoelectric sensors, such as hydrothermal synthesis [10], sol-gel synthesis [11, 12], combustion synthesis [13], and solid-state synthesis [14-17]. In this work, we have developed a method for the synthesis of a solid-state piezoelectric force sensor using sodium carbonate (Na_2CO_3), potassium carbonate (K_2CO_3), lithium carbonate (Li_2CO_3), tantalum oxide (Ta_2O_5) and niobium oxide (Nb_2O_5). FTIR, dielectric and piezoelectric characterizations were performed. An interpretation of the results of the different characterizations as well as the application of the NKLNT ceramic as a force sensor was presented.

2. Exprimental procedure

The method used for the synthesis of NKLNT ceramic is the solid route. The mixing of the raw materials was carried out in an alcoholic medium in a Teflon bowl with the help of an electric motor (attritor), in the presence of Zirconia beads to get a more homogeneous slip. After stirring for two hours, the suspension was then sieved and oven-dried for 12 hours. After that, the powder is subjected to a thermal cycle (Fig.1) using a programmable oven called calcination.

This operation aims to transform the mixture into a powder of well-defined compositions and structures. The fireclay obtained after calcination underwent a second grinding to reduce the size of the grains and increase their reactivity, this step also allows a better homogenization of the powder. Disc-shaped compacts of 11.01 mm diameter and 1.16 mm thickness were then formed using a uniaxial press at a pressure of 30 kN (Fig.2).

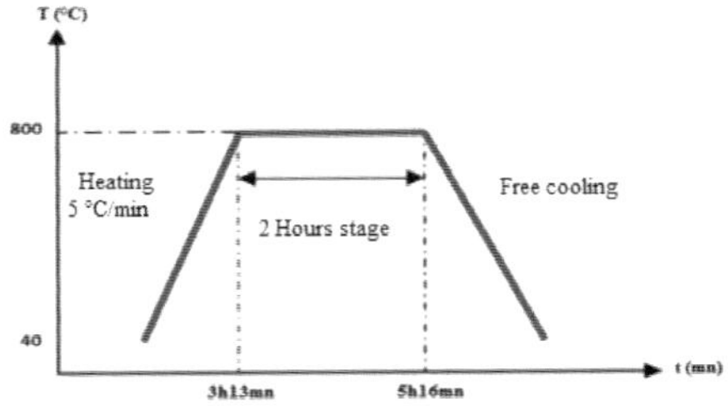

Figure 1. The thermal cycle of calcination

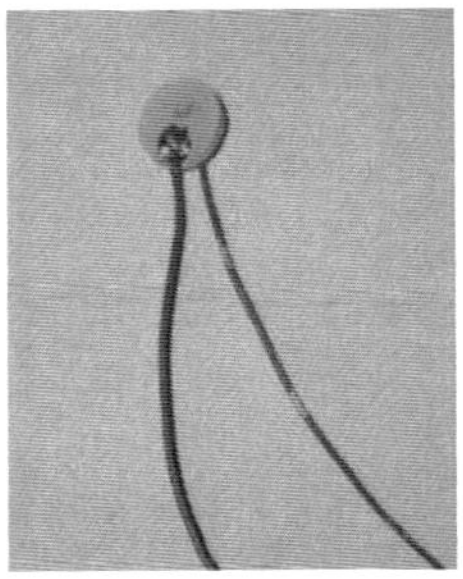

Figure 2. The elaborated sensors.

The ceramic is finally sintered at a temperature of 1110 °C for two hours with a mounting rate of 5 °C/min and natural cooling to room temperature inside the furnace. The sintering cycle is shown in Fig.2.

1) Polishing of NKLNT ceramic: The ceramics after sintering did not have the same thickness. They also have slight deformations. For this, we used the surfaces of the ceramics as well as possible and remove all the apparent porosity on the surface which allows us to decrease the dielectric losses.

2) Metallization of NKLNT ceramic: Both surfaces of the ceramic are covered with a thin layer of silver lacquer (metallization) and then dried at 100°C for 30 minutes to obtain plane capacitors. These pellets are then heat-treated at 600°C for 15 minutes to form the electrodes.

3) Polarization of NKLNT ceramic

For piezoelectric measurements, polarization of the sample took place at room temperature under a 3 kV electric field for 30 minutes.

3. Dielectric and piezoelectric characterization

Measurement of dielectric properties such as capacitance C and dielectric losses tan(δ) was performed using an impedance meter under voltage (about 1V) at different frequencies (1KHz, 10KHz, 100KHz and 1MHz) and at different temperatures (cooling from 500°C to 28°C). The permittivity is then given by the following relation:

$$\varepsilon_r = (e \times C) / (S \qquad (1)$$

- C : measured capacitance (F).
- e : distance between electrodes (m).
- S : area of electrodes (m^2).
- ε_0 : vaccum permittivity, $\varepsilon_0 = 8.85 \times 10^{-12}$ (F/m).

The evolution of the capacitance, relative permittivity, and dielectric losses as a function of the temperature at different frequencies are illustrated in Fig.3, 4 and, 5 respectively.

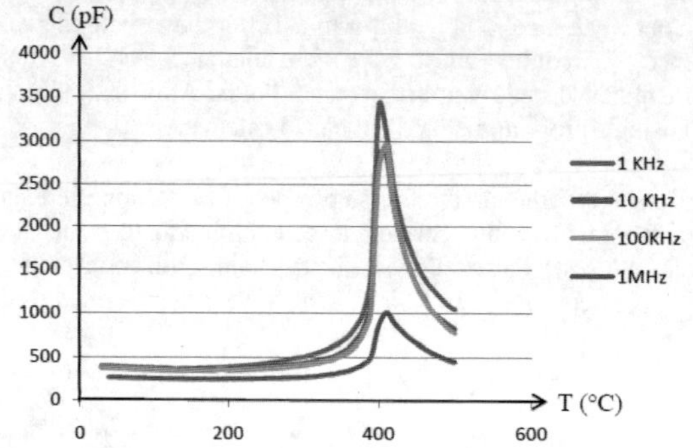

Figure 3. Evolution of capacity as a function of temperature at different frequencies.

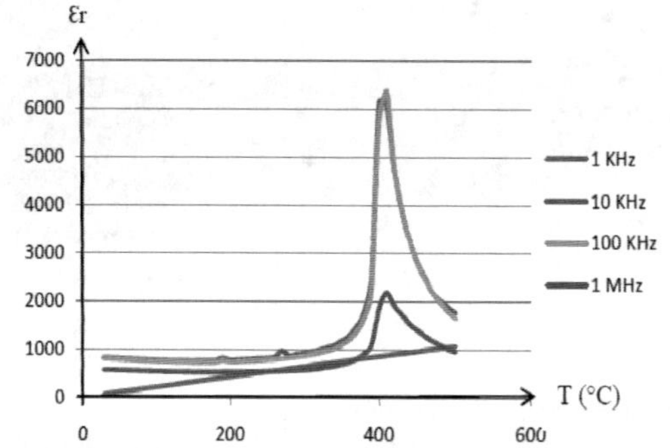

Figure 4. Evolution of relative permittivity as a function of temperature at different frequencies.

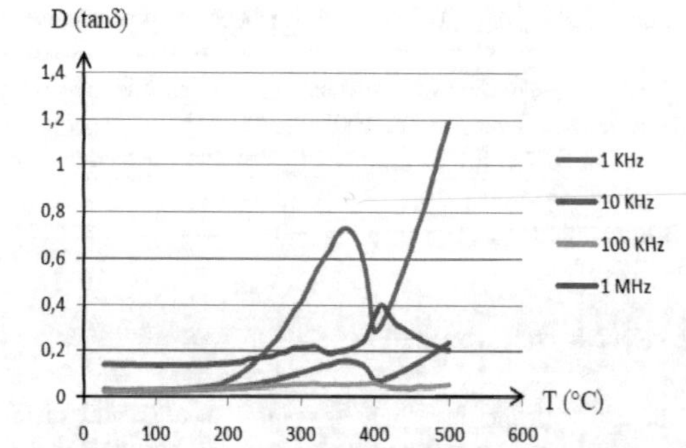

Figure 5. Evolution of dielectric losses as a function of temperature at different frequencies.

IOP Conf. Series: Materials Science and Engineering 1204 (2021) 012007 doi:10.1088/1757-899X/1204/1/012007

According to figure 4, the relative permittivity depends on the temperature. We can see that there is a transition from the orthorhombic phase to the tetragonal phase at low temperature, and the transition from the tetragonal phase to the cubic phase at a higher temperature.
The dielectric losses obtained in Figure5 are lower; they increase rapidly at higher temperature due to the rapid increase in material conductivity.
The piezoelectric constant (d_{33} coefficient = 147 pC/N) were measured using a piezoelectric meter by applying a 1N force at 100 Hz, and measuring the variation of charges Q. This coefficient is determined by the following relation:

$$\delta_{33} = Q / F. \quad (2)$$

- Q: electrical charges (pC).
- F: force (N)

4. Application of NKLNT ceramic as force sensor

a) The electrical circuit

After the polarization step, the sensor has acquired piezoelectric properties. By applying mechanical stress (vibrations), electrical charges were generated on both sides of the ceramic sensor. For this, a conditioning circuit is used to convert the generated charges into an equivalent electrical signal. This circuit is mainly composed of two parts (charge-voltage converter and amplifier stage).

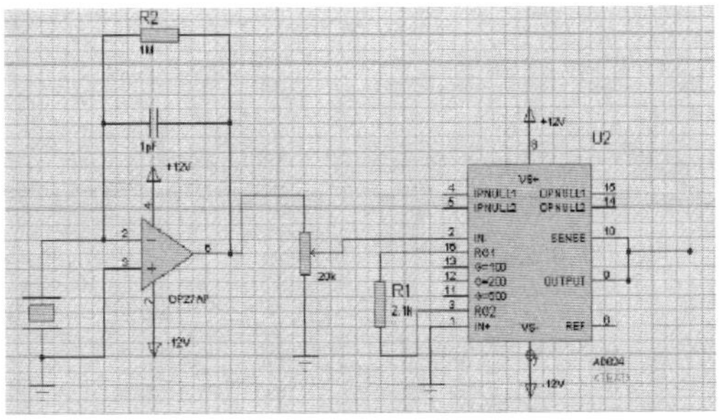

Figure 6. Electrical circuit of the conditioning system used.

b) Charge-voltage converter

The force sensor is connected to an operational amplifier which functions as an integrator to compensate for the electrical charges produced. The voltage measured at the capacitor terminals is proportional to the charges produced by the force sensor. These charges are converted into a voltage using the following equation:

$$V = - Q / C \quad (3)$$

- Q: electrical charges (pC).
- C: capacitance (F).

c) The amplification part

The output signal of the sensor is weak and noisy, so an instrumentation amplifier with high rejection and a gain of 20 was used to reduce the noise and amplify the signal.

d) Results of tests

The experimental setup for our application is shown in Fig.7.

Figure 7. The electrical circuit of the proposed application.

Fig.8. shows the output voltage of the NKLNT ceramic sensor after applying different forces (1, 3, 5, 7, and 10 N). From these results, it can be seen that the NKLNT ceramic is a technology that can fulfill the role of a (low) force sensor.

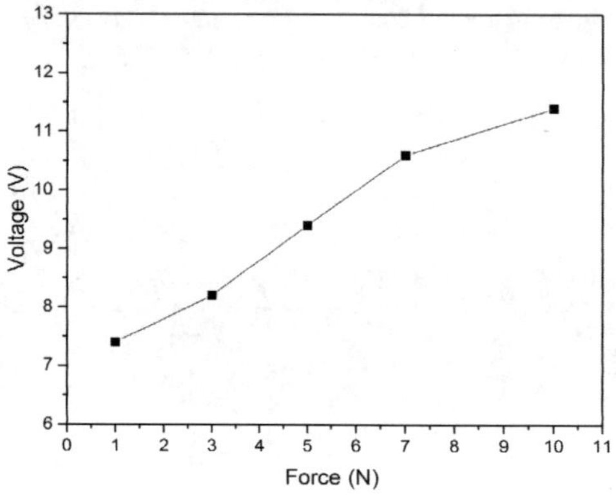

Figure 8. Output voltage of the sensor as a function of the applied force.

5. Conclusion

A force sensor based on lead-free ferroelectric particles NKLNT has been realized. The dielectric loss (tan(δ)) obtained was lower, and the dielectric constant ("ε_r") varied from 800 to 6000 at room temperature and T_C, respectively. a good piezoelectric coefficient d_{33} = 147 pC/N was measured. A relatively homogeneous microstructure and low porosity were also obtained.

These results show that the NKLNT ceramic is a technology that can fulfill the role of a force transducer for a medical application (functional hand rehabilitation).

References

[1] Li, J. F., Wang, K., Zhu, F. Y., Cheng, L. Q., & Yao, F. Z. (2013). (K, Na) NbO 3-Based Lead-Free Piezoceramics: Fundamental Aspects, Processing Technologies, and Remaining Challenges. *Journal of the American Ceramic Society*, *96*(12), 3677-3696.

[2] Kreft, H., & Jetz, W. (2007). Global patterns and determinants of vascular plant diversity. *Proceedings of the National Academy of Sciences*, *104*(14), 5925-5930.

[3] Chelli, Z., Achour, H., Saidi, M., Laghrouche, M., Chaouchi, A., Rguiti, M., ... & Courtois, C. (2021). Fabrication and characterization of PU/NKLNT/CFs based lead-free piezoelectric composite for energy harvesting application. *Polymer-Plastics Technology and Materials*, 1-13.

[4] Kras, A., Brahim, M., Porchez, T., Bouchet, C., & Claeyssen, F. (2014, June). Compact, lightweight, and efficient piezo-actuation chain for aeronautical applications. In *14th International Conference on New Actuators*.

[5] Nuffer, J., & Bein, T. (2006, October). Application of piezoelectric materials in transportation industry. In *Global symposium on innovative solutions for the advancement of the transport industry* (Vol. 4, No. 6).

[6] Choi, W. J., Jeon, Y., Jeong, J. H., Sood, R., & Kim, S. G. (2006). Energy harvesting MEMS device based on thin film piezoelectric cantilevers. *Journal of Electroceramics*, *17*(2-4), 543-548.

[7] Jeon, Y. B., Sood, R., Jeong, J. H., & Kim, S. G. (2005). MEMS power generator with transverse mode thin film PZT. *Sensors and Actuators A: Physical*, *122*(1), 16-22.

[8] Lee, B. S., Lin, S. C., Wu, W. J., Wang, X. Y., Chang, P. Z., & Lee, C. K. (2009). Piezoelectric MEMS generators fabricated with an aerosol deposition PZT thin film. *Journal of Micromechanics and Microengineering*, *19*(6), 065014.

[9] Boukhenous, S., & Attari, M. (2011). A low cost instrumentation based sensor array for ankle rehabilitation. *Chapter Book: Biomedical Engineering, trends in Electronics, Communications and Software*, 69-78.

[10] Xin, R., Ren, F., & Leng, Y. (2010). Synthesis and characterization of nano-crystalline calcium phosphates with EDTA-assisted hydrothermal method. *Materials & design*, *31*(4), 1691-1694.

[11] Barrow, D. A., Petroff, T. E., Tandon, R. P., & Sayer, M. (1997). Characterization of thick lead zirconate titanate films fabricated using a new sol gel based process. *Journal of Applied Physics*, *81*(2), 876-881.

[12] Kobayashi, M., Jen, C. K., & Levesque, D. (2006). Flexible ultrasonic transducers. *IEEE transactions on ultrasonics, ferroelectrics, and frequency control*, *53*(8), 1478-1486.

[13] Aruna, S. T., & Mukasyan, A. S. (2008). Combustion synthesis and nanomaterials. *Current opinion in solid state and materials science*, *12*(3-4), 44-50.

[14] Gabbott, P. (Ed.). (2008). *Principles and applications of thermal analysis*. John Wiley & Sons.

[15] Newman, A. W., & Byrn, S. R. (2003). Solid-state analysis of the active pharmaceutical ingredient in drug products. *Drug discovery today*, *8*(19), 898-905.

[16] Savolainen, M., Kogermann, K., Heinz, A., Aaltonen, J., Peltonen, L., Strachan, C., & Yliruusi, J. (2009). Better understanding of dissolution behaviour of amorphous drugs by in situ solid-state analysis using Raman spectroscopy. *European journal of pharmaceutics and biopharmaceutics*, *71*(1), 71-79.

[17] Saidi, M., Chaouchi, A., d'Astorg, S., Rguiti, M., & Courtois, C. (2015). Dielectric, ferroelectrics properties and impedance spectroscopy analysis of the [(Na 0.535 K 0.480) 0.966 Li 0.058](Nb 0.90 Ta 0.10) O 3-based lead-free ceramics. *Journal of Advanced Dielectrics*, *5*(01), 1550007.

Rheological behavior of the composite matrix Diglycidylether of bisphenol-A (DGEBA/ wt% blast furnace slag (BFS)

Amar Irekti[1,2], Mehena Oualit [1,2], Zohra Ykene[3] , Dorel Buncianu[4]

[1]Department of Chemistry, Faculty of Sciences, University M'Hamed Bougara de Boumerdes, Avenue de l'indépendance, 35000 Boumerdes, Algeria
[2] Laboratory of Applied Chemistry and Materials (LabCAM), University M'Hamed Bougara de Boumerdes, Avenue de l'indépendance, 35000 Boumerdes, Algeria
[3]Laboratory of Petrochemical Synthesis, University M'Hamed Bougara de Boumerdes, Avenue de l'indépendance, 35000 Boumerdes, Algeria
[4]University Politehnica Timisoara, Bdul. Mihai Viteazu No.1, Romania

a.irekti@univ-boumerdes.dz

Abstract. This paper reports experiments on the rheological behavior of DGEBA epoxy resin filled with blast furnace slag nanoparticles. The study of the variation of stress and viscosity as a function of shear rate is emphasized.
The purpose of this study is to investigate the rheological behavior of DGEBA loaded with blast furnace slag (BFS), the selected mixtures cannot exceed 10% in wt%, given the fineness of grinding which SSB= 44.5 $m^2.g^{-1}$. We tested the compositions of DGEBA/%wt blast furnace slag (BFS) at filling ratios of (10, 20, 30, 40and 50%). Viscosity and shear stress as a function of temperature and filler rate were investigated. In this study, formulations were developed to test the best compositions with favorable rheological behavior for better processing of these matrices.
Keywords: DGEBA, blast furnace slag, FTIR, rheological behavior, viscosity

1. Introduction

The search for new and improved components for coating systems requires a better understanding of the structure-property relationships of materials [1-2]. Although some properties of coating systems can be obtained experimentally, the ability to predict the properties of new coatings prior to laboratory synthesis will greatly facilitate the design of new coatings [3-4].
Resins reinforced with inorganic fillers have recently attracted considerable interest due to their potential application in a wide variety of technological fields [1-4].
The rheological behavior of resins has always played a role in understanding and controlling the processing when used in the manufacture of complex shaped composite materials [5].
The epoxy resin is usually considered as a complex fluid that exhibits various rheological behaviors under shear, most often rheofluidizing, but also rheo-thickening, or even both behaviors if the range of shear rates examined is sufficiently large. This type of resin can also show solids and threshold stress. However, the rheofluidizing behavior is by far the most frequently encountered.
In the case of using fillers as reinforcement, several authors underline the importance of the microstructure (anisotropic, isotropic) on the rheological behavior of the composite. Studies of S. Ménio (2000) [6] and Dumont et al. (2003) [7], during shear and simple compression tests on a filled epoxy matrix, showed that the elongational viscosity of these products is more than 100 times the shear viscosity, whereas for a Newtonian isotropic fluid the ratio between these two quantities is only three. R.Blanc [8] stresses the

importance of taking into account the orientation and distribution of loads imposed by the experiment carried out.
In this study, we will test the contribution of the blast furnace slag type mineral filler on the rheological behavior of the epoxy resin and the influence of temperature on the fluidity of the resin/filler system. After curing we will study the distribution and morphology of the filler in the matrix.

2. Materials and Methods

2.1. Materials

The epoxy resin [diglycidyl ether of bisphenol A (DGEBA)] with a density (ISO 758): 1.1 ± 0.05 and an epoxy equivalent weight of 171 g.eq^{-1}.
Infrared absorption spectroscopy (IR) provides very valuable information about the chemical structure of the compositions. Besides, IR is available especially for the cross-linking analysis, and it has been used for the estimation of both epoxy and hydroxyl functional groups in polymeric materials [9-11].
Aliphatic amines are well known quenchers of the aromatic hydrocarbons. It has been reported that the quenching of the fluorescence of the aromatic hydrocarbons by aliphatic amines increase with electron-donating ability of the amine groups [12]. In FT-IR spectra (Fig. 1), peaks at 890 cm^{-1} to 910 cm^{-1} are attributed to an epoxy group. A hydroxyl group is indicated by a broad band at 4000 cm^{-1}.

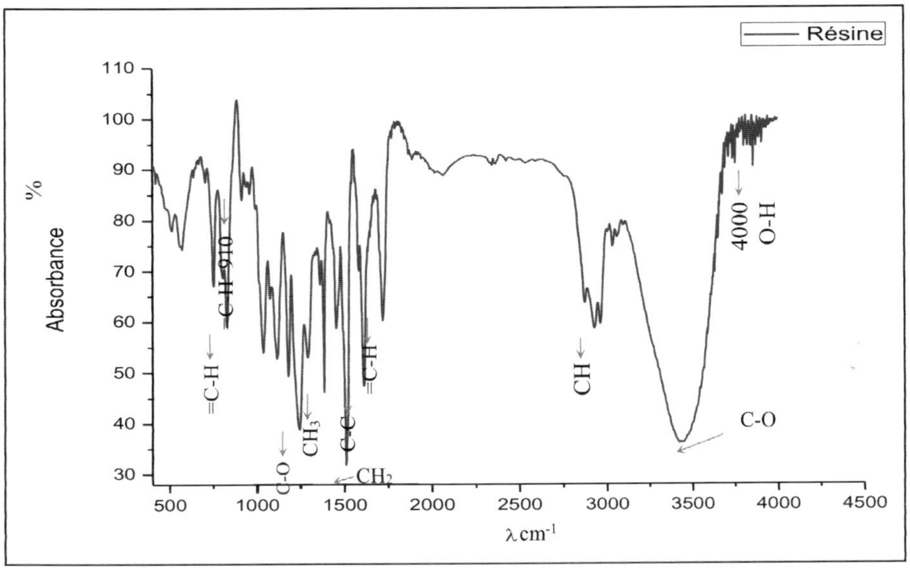

Figure 1. Fourier transform infrared spectroscopy spectra of resin DGEBA.

We use in our study the Algerian slag of the metallurgical plant of El Hadjar, Annaba. Currently the two blast furnaces of the plant produce 400 000 t/year of slag of which 20 % is crushed the rest is granulated. Blast furnace slag, or ground slag as he would like to call it, is a by-product of the manufacture of cast iron that has been cooled down by sprinkling with water, and is a hydraulic material when it is activated. It is in the form of nodules whose chemical composition includes calcium oxide in proportions of the order of 35 to 50%, silica, between 25 to 40% of lime and magnesia and up to 12% of other oxides in very small quantities (Table1).

Table 1. Typical chemical composition of Algerian blast furnace slag (BFS)

Components	SiO_2	CaO	MgO	Al_2O_3	Fe_2O_3	MgO	SO_3	Na_2O
Content (%)	51.69	28,77	4,7	7.0	1.41	5.93	0.15	0.1

IOP Conf. Series: Materials Science and Engineering 1204 (2021) 012008 doi:10.1088/1757-899X/1204/1/012008

The particle size distribution shown in Fig.2 shows that a large part of the particles has a diameter between 5 µm is between 4 µm and 5µm as well as two other peaks, one at 0.5µm and the other at 11µm. This is due to the humidity and the strong reactivity with the slag during sieving caused by the presence of CaO in large quantities. The Scanning Electron Microscope (SEM) photograph and particle size analysis shown that the filler of BFS present nodules with a porous aspect due to the water vapor trapped during the quenching, we will take into account its influence on the workability, with the high specific surface area, determined using the BET method, value of 24,1 $m^2.g^{-1}$.

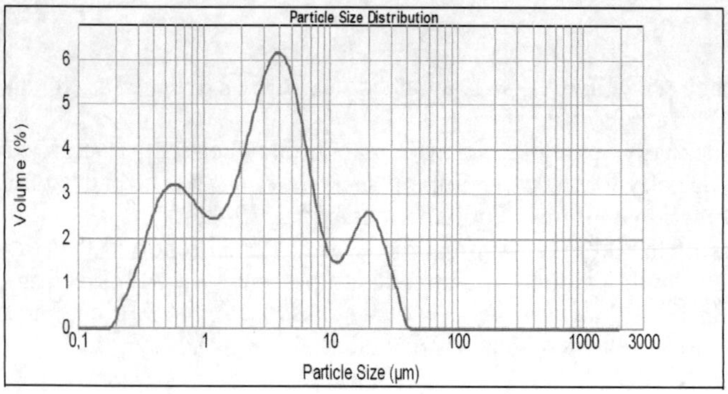

Figure 2. Particles size distribution of blast furnace slag

Figure3. SEM micrographs of blast furnace slag

2.2. Methods
The rheological behavior was studied on the Haake VT500 viscometer, which allowed us to impose a shear rate of 0 to 550 s-1 with a test temperature of 23 °C. The results of this study show the evolution of shear stress (τ) and apparent viscosity (η) as a function of shear rate.

2.3. Sample preparation
To study the rheological behavior of the elaborated matrices we used a total mass of 30 g composed of 2% by weight of diluent. The different percentages of the fillers are presented in table 2.

Table 2. Percentages for rheological tests

Diluent		2% Diluent =06g				
Control		0.6g Diluant		29.4 Resin (Element A)		
Rate	%	10	20	30	40	50
DGEBA/ %wt BFS	g	2,94	5,88	8,82	11,76	14,7

3. Results and discussions

3.1. Chemical composition

Infrared spectroscopy has been widely used for the characterization of organic compounds and many reference libraries of spectra can be found easily. Qualitative and quantitative information can be obtained by this technique, although its use in epoxy systems is rather limited due to the location and intensity of oxirane ring absorptions. Two characteristic oxirane ring absorptions are observed in the range between 4000 cm^{-1} and 400 cm^{-1}. The first, at 915^{-1} cm, is attributed to CO deformation of the oxirane group, although work by Dannenberg et al (1956) [13] has shown that this band does not correspond exclusively to this deformation, but also to another unknown process. The second band is located at about 3050 cm^{-1} and is attributed to the tension of the methylene group of the CH epoxy ring. This band is not very useful, because its intensity is low and it is also very close to strong OH absorptions, but in weak polymerization of epoxy monomers can be used as a qualitative indicator of the presence of epoxy groups.

IOP Conf. Series: Materials Science and Engineering 1204 (2021) 012008 doi:10.1088/1757-899X/1204/1/012008

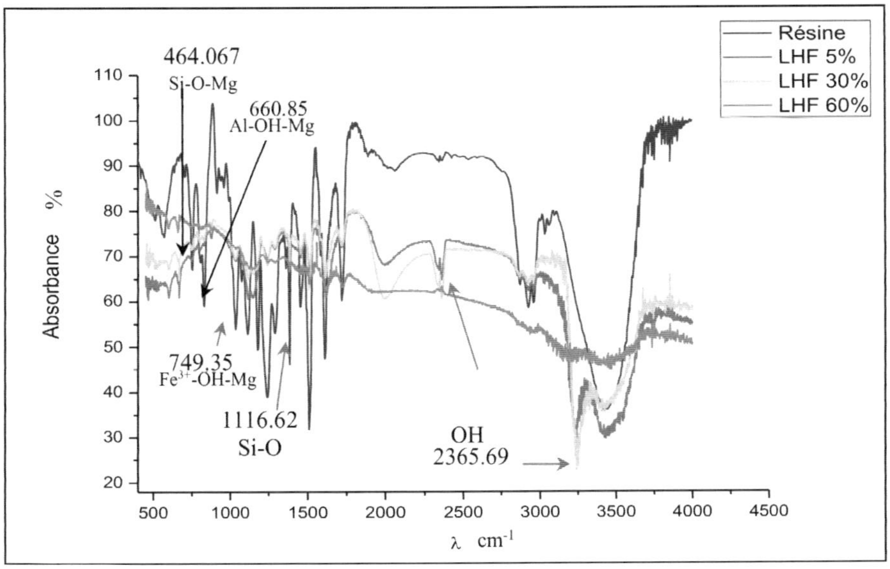

Figure 4. FTIR spectrum of the composite DGEBA/ %wt BFS

In the case of the composite with DGEBA / 35 wt% HF slag, the difference between the front view and the surface view is very visible where we find in Fig.5.a, b. a very homogeneous morphology, the particles of the slag are surrounded by the epoxy matrix and do not separate as in the case of pozzolan.

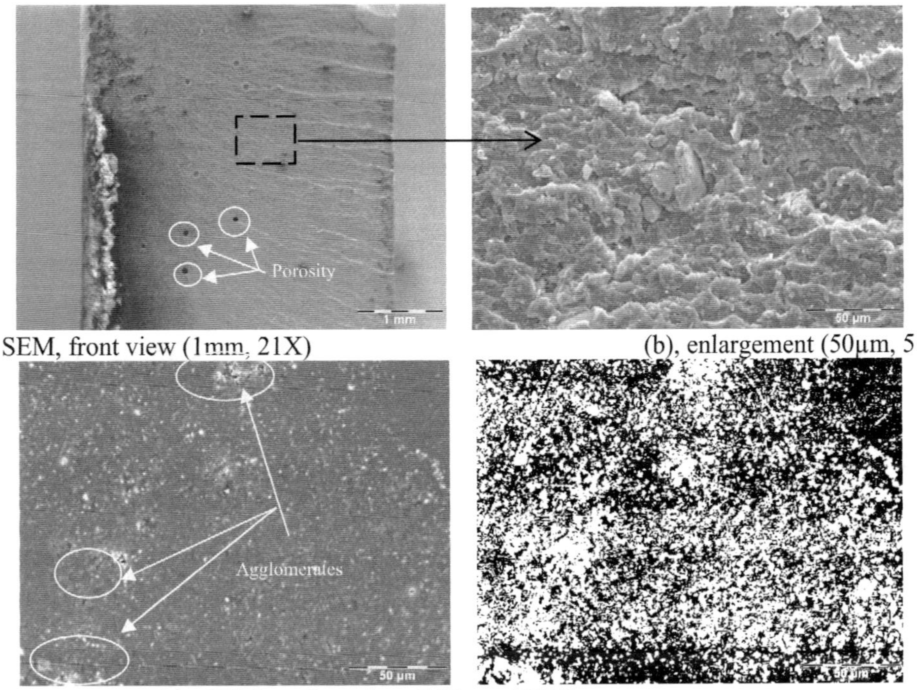

(a) : SEM, front view (1mm, 21X) (b), enlargement (50µm, 513X)

(c) : SEM, surface view (100µm, 213X)

Figure 5. SEM micrographs of the composite DGEBA/ %wt BFS

IOP Conf. Series: Materials Science and Engineering 1204 (2021) 012008 doi:10.1088/1757-899X/1204/1/012008

The surface of the sample shows more agglomeration of particles, the presence of moisture generated by the gypsum favors this type of phenomenon. These can influence the material characteristics and form microscopic defects in the form of nodules of (30μm).
According to some authors [14], the presence of water can be explained by the two-phase morphology of the epoxy networks (Figure 5): a dense, highly cross-linked phase (or micro-gel) is embedded in an inter-nodular matrix of lower cross-linking density. Steric hindrance in the highly cross-linked areas would be responsible for the inaccessibility of some sorption sites.
As in the case of the first composite, an identification of the two phases visible in fig. 5.b showed the presence of the small nodules of the mineral load rich in (Si and Ca with traces of Fe). The matrix and represented as the majority phase and given by the element C. The two phases presented are crystalline in their majority, the slag used in this study presents crystalline phases.

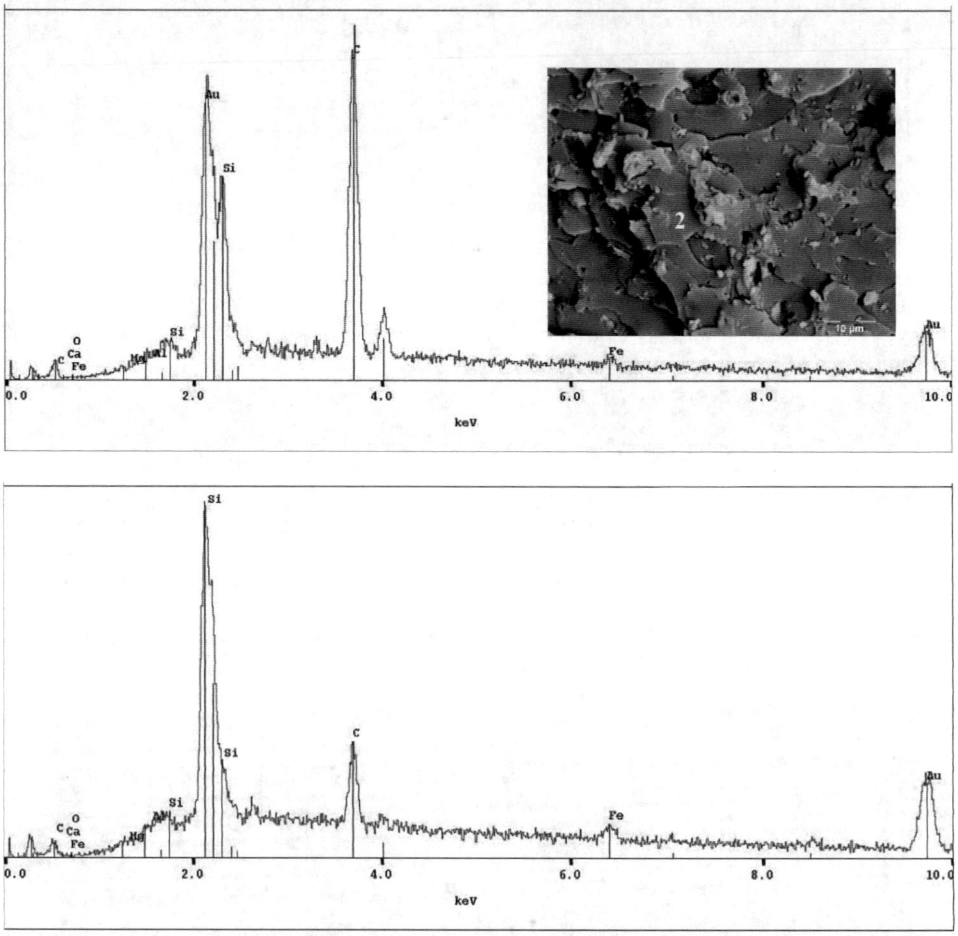

(1)

(2)

Figure 6. EDS composite DGEBA/ %wt BFS

3.2. Rheological behaviour of composite DGEBA/ %wt BFS

Rheological characterization of thermoset polymers is of great importance in understanding the fundamental nature of complex suspensions. The effect of filler distribution and size on the shaping of new epoxy-based materials requires a thorough understanding of the rheological behavior. A number of studies

have reported the rheological behavior of filled thermoplastic polymers [15-17]. But rheological studies on bisphenol-A diglycidyl ether (DGEBA) are limited. The aim of this study is to investigate the rheological behavior of DGEBA filled with silica-rich mineral powders. The viscosity and shear stress as a function of temperature and filler rate were investigated.
In this study, formulations were developed to test the best compositions with favorable rheological behavior for better processing of these matrices. The evolution of viscosity and shear stress of DGEBA resin filled at rates (10 to 50% wt%) with slag is shown in Fig. 7.
Fig.7. shows the increase in stress and viscosity with increasing percentages wt% of filler. At 50% slag the rheogram marked a very large increase in viscosity at a low shear rate.

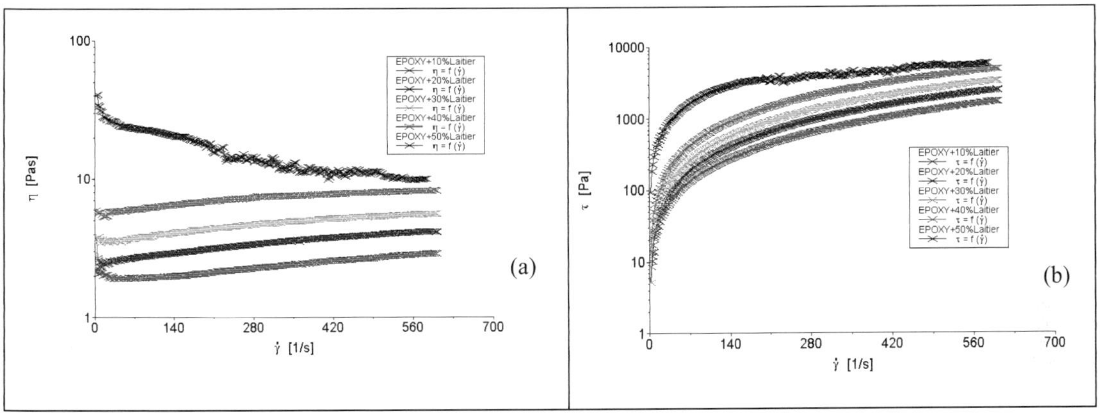

Figure 7. Rheological behavior composite DGEBA/ %wt BFS.(a) Evolution of viscosity (η) as a function of shear rate - (b) Evolution of stress (τ) as a function of shear rate

3.2.1. Rheological model

The Rheowine software of the viscometer, allowed us to identify a model and a rheological behavior of these resin blends. The obtained result confirmed that all the resin mixtures have a rheological behavior rheofluidifying and that their behavior follows the model of Ostwald de Waele [18], [19] described in the following relation (1):

$$\tau = k\,\dot{\gamma}^{n} \qquad (1)$$

Où :
τ : Shear stress;
k: Constant of consistency;
$\dot{\gamma}$: Shear rate;
n : Flow index.

The principle of the model is to consider that the flow index n and consistency K are not constants but vary with the shear rate [20].
According to this model, the value of the flow index (n) confirms the rheofluidizing behavior of these formulations. The apparent dynamic viscosity is then given by relation (2):

$$\eta_a = \frac{\tau}{\dot{\gamma}} = k\dot{\gamma}^{n-1} \qquad (2)$$

Thus, if:

- $0 < n < 1$, The fluid is rheofluidic or pseudoplastic;
- $n = 1$, it is Newtonian;
- $n > 1$, it is rheo-thickening or dilating.

The Ostwald-de Waele, Carreau-Yassuda and Cross models represent behavior without threshold stress. In the case of "threshold fluids" we add the threshold stress (τ_0). If the applied stress remains below this critical value (τ_0), the material behaves like a solid. Otherwise, the material returns to the behavior of a fluid. The two most widely used models for characterizing threshold fluids are, respectively, the Bingham (1922) model [21] and the Herschel-Bulkley (1926) model [22], where ηpl is the plastic viscosity, k is the consistency of the fluid and n is the flow index. Both of these models are able to represent the rheological behavior of many fluids above the threshold stress. We note that many other models with different fitting parameters can be found in the literature [23-24]. We ourselves have proposed a model derived from the Herschel-Bulkley model to better represent both small and large deformations for threshold fluids such as clay suspensions [25].
This model is approximate since it is only valid in a shear range whose interval depends on the fluid itself. It omits the two Newtonian plateaus (η_0 and η_∞) which are well modeled by the Carreau-Yasuda law. However, it models the behavior of polymer melts well over a wide range of the shear rate corresponding to typical values for plastic injection molding and is therefore frequently used in this field. Moreover, this model allows to compare the rheological behaviors of different fluids based on the physical indices it provides, in particular the consistency K in [Pa.s] which could summarize the rheological results of the studied mixtures. The value of K, which represents the viscosity at low shears, more precisely at $\gamma=1\ s^{-1}$, would give an indication of the cohesive state between the components of a structured fluid.

4. Conclusion

The rheological behavior study of the composite on the basis of the formulations mentioned above has highlighted the convergences of the tested mineral filler, in spite of the differences in composition and use. First, we marked the evolution of the main rheological parameters such as plastic viscosity and shear stress. It is clear from the results obtained that the mineral filler significantly increases these rheological parameters. Compared to the samples without mineral filler (DGEBA), a 50% increase in shear stress and plastic viscosity was recorded for the DGEBA/ %wt BFS composite. This can be explained by the physical filling role played by the fillers on the one hand, and on the other hand, the particle size distribution of the fillers as well as their geometrical shapes which help enormously to densify the network. The Ostwald de Waele model is repeated on all the tested compositions and confirms the rheological rheofluidifying behavior of this type of composite.

References

[1] L. Nikolic, D. Skala, V. Nikolic, J. Stamenkovic, D. Babic, and S. Ilic-Stojanovic, J. Appl. Polym. Sci., 91, 387 (2004).
[2] A. Irekti and B. Bezzazi. Rheological Study of Composite Materials Based on Thermosetting Matrix and Fillers Mineral. Key Engineering Materials, 550 (2013) 79–84, doi.org/10.4028/www.scientific.net/KEM.550.79
[3] L. Nikolic, L. Raskovic, J. Stamenkovic, and V. Nikolic, World Polym., 3, 6 (2000).
[4] Irekti, M. Oualit, H. Siahmed, D. Buncianu, F. Zibouche. Effect of black liquor from date palm on the workability and compressive strength of Portland cement and concrete. Advances in Materials Science, Vol. 19, No. 3 (61), September 2019. doi: 10.2478/adms-2019-0013
[5] Irekti Amar, et al. FTIR Analysis and Rheological Behavior of Bisphenol: A Diglycidyl Ether Resin Filled Fume-Silica " Journal of Materials Science and Engineering A 4 (11) (2014) 340-347. https://doi:10.17265/2161-6213/2014.11.004
[6] Ménio S. Optimisation du moulage par injection des composites BMC : relations écoulement/structure/état de surface, PhD thesis (2000)., Université des Sciences et Technologies de Lille, France.

[7] Dumont P. (2003), Etude des phénomènes de redistribution lors de la mise en forme de matériaux multi-constituants, PhD thesis, Université Joseph Fourier - INPG, Grenoble, France.
[8] Blanc R. Etude de l'injection des composites polyesters thermodurcissables, PhD thesis, Ecole Nationale Supérieure des Mines de Paris, France. (1988).
[9] A. Irekti, B. Bezzazi, C. Boualam, C. Aribi, and H. Dilmi, J. Mater. Sci. Eng. A 4, 340 (2014).
[10] C. Mao, C.L. Larson, and S.A. Tucker, Polycyclic Aromat. Compd., 22, 99 (2002).
[11] P. Jacues, X. Allonas, and M. Dossot, J. Photochem. Photo- biol. A: Chem. 142, 91 (2001).
[12] M. Vasilescy, Rev. Roum. Chim., 34 , 1819 (1989).
[13] H. Dannenberg, W. R. Harp,Determination of Cure and Analyses of Cured Epoxy Resins. Journal Analytical Chemistry. Chem. 28, 86(1956
[14] Kornmann H. Lindberg, L.A. Berglund. Synthesis of epoxy–clay nanocomposites: influence of the nature of the clay on structure. Polymer, Volume 42, Issue 4, February 2001, Pages 1303–1310
[15] Bost J. – Résines éthoxylique. Matières Plastiques-Chimie- Application, Technique et documentation, Paris, 1980, pp. 45- 61.
[16] Kowalska M. Carbon fibre reinforced epoxy prepregs and composites health risks aspects. Society for the Advancement of Material and Process Engineering journal (SAMPE), Q 13, 1982, pp. 13-19.
[17] Marié J.L. Dermatoses Professionnelles aux résines époxy, Document pour le Médecin du travail, INRS, Paris, ISSN 0339-6517, 2002, 306 pages.
[18] Ostwald, W.,"About the rate function of the viscosity of dispersed systems," Kolloid-Z. 36, 99–117-1925,in German.
[19] A.Irekti, B. Bezzazi, A. Smith, C. Aribi. Experimental Study of Dielectric Properties of Composite Materials Pozzolan/DGEBA. Journal Polymer Composites, Wiley, 44(82), (2015), DOI 10.1002/pc.23589
[20] Bekkour, K., and N. Kherfellah, "Linear viscoelastic behavior of bentonite-water suspensions," Appl. Rheol. 12, 234–240. 2002
[21]Bingham, E. C., "Fluidity and plasticity" McGraw-Hill (1922).
[22] Herschel, W. H., and R. Bulkley.,"Consistency measurements of rubber benzene solutions," Kolloid-Z. 39,291–300 _1926_, in German
[23] Roberts, G. P., H. A. Barnes and P. Carew, "Modelling the flow behaviour of very shear-thinning liquids" 56, 5617-5623 (2001).
[24] Mendes, P. R. S. and E. S. S. Dutra, "Viscosity Function for Yield-Stress Liquids" Appl. Rheol. 14, 296-302 (2004).
[25] Benchabane, A. and K. Bekkour. Couplage surfactant-polymère dans une suspension argileuse. 40ème Colloque annuel du GFR. Nice 2005b, p. 65-68.

AUTHOR INDEX